LONDON MATHEMATICAL SOCIETY LECTURE NOTE SERIES

Managing Editor: Professor M. Reid, Mathematics Institute,
University of Warwick, Coventry CV4 7AL, United Kingdom

The titles below are available from booksellers, or from Cambridge University Press at
http://www.cambridge.org/mathematics

London Mathematical Society Lecture Note Series: 439

Evolution Equations:
Long Time Behavior and Control

Edited by

KAÏS AMMARI
University of Monsatir, Tunisia

STÉPHANE GERBI
University Savoie Mont Blanc, Chambéry, France

CAMBRIDGE
UNIVERSITY PRESS

CAMBRIDGE
UNIVERSITY PRESS

Shaftesbury Road, Cambridge CB2 8EA, United Kingdom

One Liberty Plaza, 20th Floor, New York, NY 10006, USA

477 Williamstown Road, Port Melbourne, VIC 3207, Australia

314–321, 3rd Floor, Plot 3, Splendor Forum, Jasola District Centre, New Delhi – 110025, India

103 Penang Road, #05–06/07, Visioncrest Commercial, Singapore 238467

Cambridge University Press is part of Cambridge University Press & Assessment, a department of the University of Cambridge.

We share the University's mission to contribute to society through the pursuit of education, learning and research at the highest international levels of excellence.

www.cambridge.org
Information on this title: www.cambridge.org/9781108412308

DOI: 10.1017/9781108412308

First published 2018

A catalogue record for this publication is available from the British Library

ISBN   978-1-108-41230-8   Paperback

Contents

Preface

This volume constitutes the proceedings of the summer school MIS 2015, "Mathematics In Savoie 2015," whose theme was: "Evolution Equations: long time behavior and control."

This summer school was held at the University Savoie Mont Blanc, Chambéry in the period June 15–18, 2015 (see http://lama.univ-savoie.fr/ MIS2015 for details). It was organized by Kaïs Ammari, UR Analysis and Control of PDE, University of Monastir, Tunisia, and Stéphane Gerbi, Laboratoire de Mathématiques, University Savoie Mont Blanc, France.

The summer school consisted of two mini-courses in the morning while the afternoons were devoted to various contributions on the theme.

The first mini-course was held by Farid Ammar-Khodja, University of Franche-Comté, France. The topic was: "Controllability of parabolic systems: the moment method." This recent point of view on the controllability of parabolic systems permits to overview the moment method for parabolic equations. This course constitutes the first part of this volume.

The second part of this volume is devoted to the second mini-course which was held by Emmanuel Trélat, UPMC, Paris. The topic was "Stabilization of semilinear PDEs, and uniform decay under discretization." This course was devoted to the numerical stabilization and control of partial differential equations and more specifically it addresses the problem of the construction of numerical feedback control that will preserve the theoretical rate of decay.

Several of the speakers agreed to write review papers related to their contributions to the summer school, while others have written more traditional research papers, which constitute the last part of this volume.

We believe that this volume therefore provides an accessible summary of a wide range of active research topics, along with some exciting new results, and we hope that it will prove a useful resource for both graduate students new to the area and to more established researchers.

The summer school brought together internationally leading researchers from the community of control theory and young researchers who came from all around the world. The organizers' intention was to provide a wide angle snapshot of this exciting and fast moving area and facilitate the exchange of ideas on recent advances in its various aspects. The numerous formal, informal, and sometimes lively discussions that resulted from this interaction were for us a sign that we achieved something in the direction of fulfilling this aim.

Our second aim was to ensure that the diffusion of these recent results was not limited to established researchers in the area who were present at the summer school, but also available to newcomers and more junior members of the research community. This was reflected by the presence of many unfamiliar and/or young faces in the audience. The present proceedings should hopefully complete the fulfillment of our second aim.

This summer school would not have materialized without the help and support of the following institutions.

We are very grateful to the CNRS (Centre National de la Recherche Scientifique), the University Savoie Mont Blanc; La Région Auvergne-Rhône-Alpes; the GDRI LEM2I: "Laboratoire Euro-Maghrébin de Mathématiques et leurs Interactions;" the GDR MACS: "Modelisation, Analyse et Conduite des Systèmes dynamiques;" the GDR EDP: "Equations aux dérivées partielles;" the GDRE CONEDP: "Control of Partial Differential Equations;" the MaiMoSine: "Maison de la Modélisation et de la Simulation, Nanosciences et Environnement;" and the PERSYVAL-lab: "PERvasive SYstems and ALgorithms" for their financial support without which this summer school would not be accessible without fees.

Finally we would like to thank all the participants of the summer school who have made this event a success, the contributors to these proceedings, and the reviewers for their hard work.

Kaïs Ammari and Stéphane Gerbi
Chambéry, July 07, 2017

List of Contributors
Present at the Summer School

Farid Ammar Khodja
University and ESPE of Franche-Comté
16, Route de Gray, 25030 Besançon Cedex, France
fammarkh@univ-fcomte.fr

Carlos Castro
Department of Mathematics and Information
ETSI Roads, Canals, and Ports
Technical University of Madrid
Ciudad Universitaria
28040 Madrid, Spain
carlos.castro@upm.es

Taoufik Hmidi
University of Rennes1
Campus de Beaulieu, IRMAR
263, Avenue du Général Leclerc
35042 Rennes, France
thmidi@univ-rennes1.fr

Arnaud Münch
Blaise Pascal University
Laboratoire de Mathématiques, UMR CNRS 6620
Clermont-Ferrand, France
arnaud.munch@math.univ-bpclermont.fr

Serge Nicaise
University of Valenciennes and of Hainaut Cambrésis
Le Mont Houy
59313 Valenciennes Cedex 9, France
snicaise@univ-valenciennes.fr

Cristina Pignotti
Department of Engineering and Computer Science and Mathematics
Via Vetoio, Loc. Coppito
67010 L'Aquila, Italy
pignotti@univaq.it

Reinhard Racke
Department of Mathematics and Statistics
University of Konstanz
Fach D 187, 78457 Konstanz, Germany
reinhard.racke@uni-konstanz.de

Lionel Rosier
Automatic Control and Systems Center
MINES ParisTech
60 Bd Saint-Michel, 75272 Paris Cedex, France
lionel.rosier@mines-paristech.fr

Armen Shirikyan
Department of Mathematics
Université de Cergy-Pontoise
Site de Saint Martin
2, Avenue Adolphe Chauvin
95302 Cergy-Pontoise Cedex, France
Armen.Shirikyan@u-cergy.fr

Emmanuel Trélat
University of Pierre et Marie Curie (Paris 6)
Laboratoire Jacques-Louis Lions
CNRS, UMR 7598
4 Place Jussieu, BC 187
75252 Paris Cedex 05, France
emmanuel.trelat@upmc.fr

1

Controllability of Parabolic Systems: The Moment Method

FARID AMMAR-KHODJA

Abstract

We give some recent controllability results of linear hyperbolic systems and we will apply them to solve some nonlinear control problems.

Mathematics Subject Classification 2010. 93B05, 93B07, 93C20, 93C05, 35K40

Key words and phrases. Parabolic systems, null controllability, moment method

Contents

1

1.1 Introduction

The main goal of these notes is to give a review of results relating to controllability issues for some parabolic systems obtained via the *moment method*. We will follow Fattorini and Russell who, in the 1970s, solved controllability problems for scalar parabolic equations (see [10, 11]). This method is very efficient in the one-dimensional space setting. But it has also been used to prove the boundary null-controllability of the heat equation for particular geometries of the space domain (disks, parallelipepidons, etc.).

At the beginning of the 1990s, Fursikov and Imanuvilov [12] solved the null-controllability problem for a general second-order parabolic equation. They did this by proving a global Carleman inequality for solutions of quite general parabolic equations. This Carleman inequality implies observability inequality and thus controllability of the corresponding parabolic equation when the control function acts on an arbitrary open subset of the space domain or on an arbitrary relatively open subset of its boundary. At the same time, Lebeau and Robbiano [16] also proved the null-controllability of the heat equation with constant coefficients. Their method of proof is less general than that of Fursikov–Imanuvilov when dealing with parabolic equations but it generalizes to abstract diagonal systems.

Since then, a huge literature has been devoted to solving control problems by a systematic use of Carleman estimates: Stokes and Navier–Stokes equations, Burger's equations, etc. But as usual in mathematics, any powerful tool or method has its limitations. These appeared in particular when dealing with parabolic systems. It is one of the goals of these notes to explain these limits.

1.2 Parabolic Systems and Controllability Concepts

Consider the following system:

$$\begin{cases} (\partial_t - D\Delta - A)y = Bu1_\omega, & Q_T := (0, T) \times \Omega, \\ y = Cv1_{\Gamma_0}, & \Sigma_T := (0, T) \times \partial\Omega, \qquad (1.1) \\ y(0, \cdot) = y^0, & \Omega, \end{cases}$$

where

- $\Omega \subset \mathbb{R}^N$ is a smooth bounded domain, $\omega \subset \Omega$ is an open set, $\Gamma_0 \subset \partial\Omega$ is a relatively open subset;

- $D = \text{diag}(d_1, ..., d_n)$, $A = (a_{ij})_{1 \le i,j \le N} \in L^\infty(Q_T; \mathcal{L}(\mathbb{R}^n))$,
- $B = (b_{ij})$, $C = (c_{ij}) \in L^\infty(Q_T; \mathcal{L}(\mathbb{R}^m, \mathbb{R}^n))$: control matrices.

Definition 1.1 System (1.1) is approximately controllable at time $T > 0$ if for all $\varepsilon > 0$, for all $(y^0, y^1) \in X \times X$, there exists $(u, v) \in L^2(Q_T) \times L^2(\Sigma_T)$ such that $\|y(T) - y^1\|_X \le \varepsilon$.

System (1.1) is null-controllable at time $T > 0$ if for all $y^0 \in X$, there exists $(u, v) \in L^2(Q_T) \times L^2(\Sigma_T)$ such that $y(T) = 0$ in Ω.

Here X is a space where the system (1.1) is well-posed. For example, when $C = 0$ (*distributed control*), it is enough to work with $X = L^2(\Omega; \mathbb{R}^n)$. In this case, variational methods should prove that for $(y^0, u) \in X \times L^2(Q_T; \mathbb{R}^m)$, system (1.1) admits a unique solution

$$y \in C([0, T]; X) \cap L^2(0, T; H_0^1(\Omega, \mathbb{R}^n)).$$

When $B = 0$ and $C \ne 0$ (*boundary control*), a suitable space is $X = H^{-1}(\Omega; \mathbb{R}^n)$. The transposition method proves that for $(y^0, u) \in X \times L^2(\Sigma_T; \mathbb{R}^m)$, system (1.1) admits a unique solution

$$y \in C([0, T]; X) \cap L^2(Q_T; \mathbb{R}^n).$$

The previous two controllability concepts have dual equivalent concepts. Introduce the backward adjoint system:

$$\begin{cases} (\partial_t + D\Delta + A^*)\varphi = 0, & \text{in} \quad Q_T, \\ \varphi = 0, & \text{on} \quad \Sigma_T, \\ \varphi(T) = \varphi^0, & \text{in} \quad \Omega. \end{cases} \tag{1.2}$$

If $\varphi^0 \in L^2(\Omega, \mathbb{R}^n)$ (resp. $\varphi^0 \in H_0^1(\Omega, \mathbb{R}^n)$) then there exists a unique solution φ to (1.2) such that:

$$\varphi \in C(0, T; L^2(\Omega, \mathbb{R}^n)) \cap L^2(0, T; H_0^1(\Omega, \mathbb{R}^n)),$$

$$\left(\text{resp. } \varphi \in C(0, T; H_0^1(\Omega, \mathbb{R}^n)) \cap L^2(0, T; H^2 \cap H_0^1(\Omega, \mathbb{R}^n))\right).$$

The following characterizations have been known for a long time and their proof can be found in [9] for instance.

Proposition 1.2
- **Assume that $C = 0$ (distributed control)**

System (1.1) is approximately controllable if, and only if, for any $\varphi^0 \in L^2(\Omega, \mathbb{R}^n)$ the associated solution to (1.2) satisfies the property:

$$B^*\varphi = 0 \quad \text{in} \quad (0, T) \times \omega \Rightarrow \varphi = 0 \quad \text{in} \quad Q_T. \tag{1.3}$$

System (1.1) is null-controllable if, and only if, there exists $C = C_T > 0$ such that for any solution to (1.2)

$$\|\varphi(0)\|^2_{L^2(\Omega,\mathbb{R}^n)} \leq C \int_0^T \int_\omega |B^*\varphi|^2 \, dxdt. \tag{1.4}$$

- **Assume $B = 0$ (boundary control)**
  System (1.1) is approximately controllable if, and only if, for any $\varphi^0 \in H_0^1(\Omega, \mathbb{R}^n)$ the associated solution to (1.2) satisfies the property:

$$C^* \frac{\partial \varphi}{\partial \nu} = 0 \quad \text{in} \quad (0, T) \times \Gamma_0 \Rightarrow \varphi = 0 \quad \text{in} \quad Q_T. \tag{1.5}$$

System (1.1) is null-controllable if, and only if, there exists $C = C_T > 0$ such that for any solution to (1.2)

$$\|\varphi(0)\|^2_{H_0^1(\Omega,\mathbb{R}^n)} \leq C \int_0^T \int_{\Gamma_0} \left| C^* \frac{\partial \varphi}{\partial \nu} \right|^2 \, dxdt.$$

1.3 Controllability Results for the Scalar Case: The Carleman Inequality

We describe in this section known controllability results for the scalar parabolic equation and give (without proof) the general form of the Carleman inequality proved in [12].

Theorem 1.3 *The problem*

$$\begin{cases} (\partial_t - \Delta - a) y = u1_\omega, & Q_T := (0, T) \times \Omega, \\ y = 0, & \Sigma_T := (0, T) \times \partial\Omega, \\ y(0, \cdot) = y^0, & \Omega, \end{cases} \tag{1.6}$$

is null and approximately controllable in $\mathbb{X} = L^2(\Omega)$ for any open set $\omega \subset \Omega$, provided that $a \in L^\infty(Q_T)$.

As a consequence, the problem

$$\begin{cases} (\partial_t - \Delta - a) y = 0, & Q_T := (0, T) \times \Omega, \\ y = v1_{\Gamma_0}, & \Sigma_T := (0, T) \times \partial\Omega, \\ y(0, \cdot) = y^0, & \Omega, \end{cases} \tag{1.7}$$

is null and approximately controllable in $\mathbb{X} = H^{-1}(\Omega)$ for any relatively open set $\Gamma_0 \subset \partial\Omega$.

To prove this result, let $\beta_0 \in C^2(\overline{\Omega})$ and $s \in \mathbb{R}$ a parameter. Introduce the functions

$$\eta(t,x) := s \frac{\beta_0(x)}{t(T-t)}, \qquad (t,x) \in Q_T,$$

$$\rho(t) := \frac{s}{t(T-t)}, \qquad (t,x) \in Q_T$$

and the functional

$$I(\tau,\varphi) = \int_{Q_T} \rho^{\tau-1} e^{-2\eta} \left(|\varphi_t|^2 + |\Delta\varphi|^2 + \rho^2 |\nabla\varphi|^2 + \rho^4 |\varphi|^2 \right).$$

Theorem 1.4 (Carleman inequality) *There exist a positive function $\beta_0 \in C^2(\overline{\Omega})$, $s_0 > 0$ and $C > 0$ such that $\forall s \geq s_0$ and $\forall \tau \in \mathbb{R}$:*

$$I(\tau,\varphi) \leq C \left(\int_{Q_T} \rho^{\tau} e^{-2\eta} |\varphi_t \pm c\Delta\varphi|^2 + \int_0^T \int_\omega \rho^{\tau+3} e^{-2\eta} |\varphi|^2 \right), \qquad (1.8)$$

for any function φ satisfying $\varphi = 0$ on Σ_T and for which the right-hand side is defined.

More detailed information about the function β_0 can be found in [12].

Let us see how this inequality is applied to prove null and approximate controllability of a system (1.6). Consider the associated backward adjoint system:

$$\begin{cases} (\partial_t + \Delta + a)\,\varphi = 0, & Q_T := (0,T) \times \Omega, \\ \varphi = 0, & \Sigma_T := (0,T) \times \partial\Omega, \\ \varphi(T,\cdot) = \varphi^0, & \Omega. \end{cases} \qquad (1.9)$$

From Theorem 1.4, for any $\varphi^0 \in L^2(\Omega)$, the solution of (1.9) satisfies (1.8) which, in particular gives the estimate:

$$\int_{Q_T} \rho^{\tau+3} e^{-2\eta} |\varphi|^2 \leq C \left(\int_{Q_T} \rho^{\tau} e^{-2\eta} |a\varphi|^2 + \int_0^T \int_\omega \rho^{\tau+3} e^{-2\eta} |\varphi|^2 \right).$$

Since

$$\int_{Q_T} \rho^{\tau} e^{-2\eta} |a\varphi|^2 \leq \|a\|_\infty^2 \int_{Q_T} \rho^{\tau} e^{-2\eta} |\varphi|^2$$

it appears that

$$\int_{Q_T} \rho^{\tau} \left(\rho^3 - \|a\|_\infty^2 \right) e^{-2\eta} |\varphi|^2 \leq C \int_0^T \int_\omega \rho^{\tau+3} e^{-2\eta} |\varphi|^2. \qquad (1.10)$$

But, for $s > 0$, we have $\rho^3 \geq \frac{4^3 s^3}{T^6}$ and taking $s \geq \frac{T^2}{2^{5/3}} \|a\|_\infty^{2/3}$, we see that $\rho^3 - \|a\|_\infty^2 > 0$ on $(0,T)$. With this choice of the parameter s, the approximate controllability property is readily implied by (1.10).

To prove the null-controllability property, something more has to be done. According to (1.4), we have to deduce from (1.10) that

$$\int_\Omega |\varphi(0,x)|^2 \, dx \le C_T \int_0^T \int_\omega |\varphi|^2 ,$$

for any solution of (1.9). After noting that $e^{-2\eta} \ge e^{-2s\overline{\beta}_0\rho}$ (here $\overline{\beta}_0 = \max_{\overline{\Omega}} \beta_0$), the other argument is that there exists $\alpha = \alpha(\|a\|_\infty)$ such that the function $t \mapsto E(t) := e^{\alpha t} \int_\Omega \varphi^2$ is increasing on $(0, T)$ (this is quite easy: it suffices to compute $E'(t)$, to use the equation satisfied by φ and to choose α in such a way that $E'(t) \le 0$ for $t \in (0, T)$). Using this, we get

$$\int_{Q_T} \rho^\tau \Big(\rho^3 - \|a\|_\infty^2\Big) e^{-2\eta} |\varphi|^2 \ge \int_0^T \rho^\tau \Big(\rho^3 - \|a\|_\infty^2\Big) e^{-2s\overline{\beta}_0\rho - \alpha t} \Big(e^{\alpha t}\int_\Omega |\varphi|^2 \, dx\Big) dt$$

$$\ge \int_0^T \rho^\tau \Big(\rho^3 - \|a\|_\infty^2\Big) e^{-2s\overline{\beta}_0\rho - \alpha t} dt \int_\Omega |\varphi(0,x)|^2 \, dx$$

$$\ge m_T \int_\Omega |\varphi(0,x)|^2 \, dx.$$

On the other hand, there exists $c_T > 0$ such that

$$\int_0^T \int_\omega \rho^{\tau+3} e^{-2\eta} |\varphi|^2 \le c_T \int_0^T \int_\omega |\varphi|^2 .$$

We arrive to: Ω

$$\int_\Omega |\varphi(0,x)|^2 \, dx \le C_T \int_0^T \int_\omega |\varphi|^2 ,$$

which is exactly the observability inequality (1.4). This proves the distributed null-controllability.

Due to this distributed null-controllability property holding true for any open subset $\omega \subset \Omega$, it allows to deduce the boundary controllability result for an arbitrary relatively open subset $\Gamma_0 \subset \partial\Omega$. Here is the (heuristic) proof. Let $\Omega' \supset \Omega$ another smooth bounded domain such that $\Omega' = \Omega \cup \Omega_0$ with $\Omega \cap \Omega_0 = \varnothing$ and $\overline{\Omega} \cap \overline{\Omega_0} \subset \Gamma_0$. By the previous result, the problem (1.6) is null-controllable on $Q'_T = (0, T) \times \Omega'$ with any $\omega \subset \Omega_0$. The restriction to $Q_T = (0, T) \times \Omega$ of a controlled solution on Q'_T is a controlled solution of system (1.7) (and the control function is just the Dirichlet trace of this controlled solution to $(0, T) \times \Gamma_0$.

Remark 1.5 Note that the Carleman inequality (1.8) allows to prove both null and approximate controllability.

1.4 First Application to a Parabolic System

Consider the 2×2 parabolic system:

$$\begin{cases} (\partial_t - \Delta)y_1 = a_{11}y_1 + a_{12}y_2 & \\ (\partial_t - d\Delta)y_2 = a_{21}y_1 + a_{22}y_2 + u1_\omega, & Q_T, \\ y = (y_1, y_2) = 0, & \Sigma_T, \\ y(0, \cdot) = y^0, & \Omega, \end{cases} \tag{1.11}$$

where $a_{ij} \in L^\infty(Q_T)$. The following result is proved in [2] and in a most general version in [13].

Theorem 1.6 *If there exists $\omega_0 \subset \omega$ such that $a_{12} \geq \sigma > 0$ on $(0, T) \times \omega_0$ then system (1.11) is null and approximately controllable for any $d > 0$.*

The proof of this result uses Carleman inequalities for scalar parabolic equations (see Theorem 1.4) applied to each equation of the backward adjoint system:

$$\begin{cases} -(\partial_t + \Delta)\varphi_1 = a_{11}\varphi_1 + a_{21}\varphi_2 & \\ -(\partial_t + \Delta)\varphi_2 = a_{12}\varphi_1 + a_{22}\varphi_2, & Q_T, \\ \varphi = (\varphi_1, \varphi_2) = 0, & \Sigma_T, \\ \varphi(0, \cdot) = \varphi^0, & \Omega. \end{cases} \tag{1.12}$$

The assumption $a_{12} \geq \sigma > 0$ on $(0, T) \times \omega_0$ is used to get an estimate of the L^2- norm of φ_1 on $(0, T) \times \omega_0$ using the second equation in (1.12). For more precise details, see [2, 13].

Natural questions arise at this level:

- What happens if

$$\text{supp}(a_{12}) \cap \omega = \varnothing?$$

The technique of proof used for the previous theorem cannot be extended to this case. It seems that Carleman estimates cannot treat this kind of situation.
- What happens for the boundary control system:

$$\begin{cases} (\partial_t - \Delta)y_1 = a_{11}y_1 + a_{12}y_2 & \\ (\partial_t - d\Delta)y_2 = a_{21}y_1 + a_{22}y_2, & Q_T, \\ y = \begin{pmatrix} y_1 \\ y_2 \end{pmatrix} = \begin{pmatrix} 0 \\ 1 \end{pmatrix} 1_{\Gamma_0}v, & \Sigma_T, \\ y(0, \cdot) = y^0, & \Omega, \end{cases}$$

where Γ_0 is a relatively open subset of $\partial\Omega$?

There exist only partial answers to these two questions: even in the one-dimensional space case. In any space dimension, the single result is the one proved by Alabau-Boussouira and Léautaud in [1]. They considered the special system

$$
\begin{cases}
(\partial_t - \Delta)y_1 = ay_1 + by_2 & \\
(\partial_t - d\Delta)y_2 = \delta by_1 + ay_2 + u1_\omega, & Q_T, \\
y = (y_1, y_2) = 0, & \Sigma_T, \\
y(0, \cdot) = y^0, & \Omega,
\end{cases}
\tag{1.13}
$$

and proved.

Theorem 1.7 [1] *Let $b \geq 0$ on Ω. Assume that there exists $b_0 > 0$ and $\omega_b := supp(b) \subset \Omega$ satisfying the Geometric Control Condition (GCC) (see [6]) with $b \geq b_0$ in ω_b. Assume that ω also satisfies GCC. Then there exists $\delta_0 > 0$ such that if $0 < \sqrt{\delta} \|b\|_{L^\infty(\Omega)} \leq \delta_0$, System (1.13) is null controllable at any positive time T.*

Carleman's inequalities are not used in the proof of this result. It is obtained as a consequence of the controllability of the corresponding hyperbolic system of two wave equations and the transmutation method.

In the forthcoming sections, we will study the one-dimensional version of system (1.11) by means of the moment method.

1.5 The Moment Method

1.5.1 Presentation: Example 1

We present in this section the moment method through the study of the null controllability issue for the scalar one-dimensional heat equation:

$$
\begin{cases}
y' - y_{xx} = f(x)u(t), & Q_T = (0, T) \times (0, \pi) \\
y_{|x=0,\pi} = 0, & (0, T) \\
y_{|t=0} = y^0 & (0, \pi).
\end{cases}
\tag{1.14}
$$

Here the constraint is that the control has separate variables: $f \in L^2(0, \pi)$ and $u \in L^2(0, T)$.

If $\varphi_k(x) = \sqrt{\frac{2}{\pi}} \sin(kx)$, then $\{\varphi_k\}_{k \geq 1}$ is an orthonormal basis of $L^2(0, \pi)$. We look for a solution in the form

$$
y(t, x) = \sum_{k \geq 1} y_k(t)\varphi_k(x).
$$

Set

$$f(x) = \sum_{k\geq 1} f_k \sin(kx), \quad y^0 = \sum_{k\geq 1} y_k^0 \sin(kx).$$

Then y is a solution if, and only if,

$$\begin{cases} y_k' = -k^2 y_k + f_k u(t), & (0,T) \\ y_{k|t=0} = y_k^0, \end{cases} \quad \forall k \geq 1,$$

i.e.

$$y_k(t) = e^{-k^2 t} y_k^0 + f_k \int_0^t e^{-k^2(t-s)} u(s)\, ds, \quad \forall k \geq 1.$$

Therefore, there exists a control function $u \in L^2(0,T)$ such that the solution satisfies $y(T,x) = 0$ for any $x \in (0,\pi)$ if, and only if, there exists $u \in L^2(0,T)$ such that:

$$f_k \int_0^T e^{-k^2(T-s)} u(s)\, ds = -e^{-k^2 T} y_k^0, \quad \forall k \geq 1.$$

After a change of variable in the integral, we arrive to the reduction of the null-controllability issue to the problem $(v(t) = u(T-t))$

$$\begin{cases} \text{Find } v \in L^2(0,T): \\ \boxed{f_k \int_0^T e^{-k^2 t} v(t)\, dt = -e^{-k^2 T} y_k^0, \quad k \geq 1.} \end{cases} \tag{1.15}$$

This is a **moment problem** in $L^2(0,T)$ with respect to the family $\{e^{-k^2 t}\}_{k\geq 1}$.

A necessary condition for the existence of a solution *for any* $y^0 \in L^2(0,\pi)$ is:

$$f_k \neq 0, \quad k \geq 1.$$

If $\{e^{-k^2 t}\}_{k\geq 1}$ admits a **biorthogonal family** $\{q_k\}_{k\geq 1}$ in $L^2(0,T)$, i.e. a family $\{q_k\}_{k\geq 1}$ such that

$$\int_0^T e^{-k^2 t} q_\ell(t)\, dt = \delta_{k\ell}, \quad k,\ell \geq 1,$$

then a formal solution is

$$v(t) = -\sum_{k\geq 1} \frac{e^{-k^2 T}}{f_k} y_k^0 q_k.$$

The question is then: $v \in L^2(0,T)$?

The next subsection is devoted to proving the existence of this biorthogonal family $\{q_k\}_{k\geq 1} \subset L^2(0, T)$ and to the estimate of $\|q_k\|_{L^2(0,T)}$ as k tends to ∞ (in order to prove that $v \in L^2(0, T)$).

1.5.2 Generalization of the Moment Problem

Let $\{\lambda_k\} \subset \mathbb{R}$ such that

$$0 < \lambda_1 < \lambda_2 < \cdots < \lambda_k < \cdots, \quad \lim_{k\to\infty} \lambda_k = \infty.$$

Let $\{m_k\}_{k\geq 1} \in \ell^2$ and consider the moment problem:

$$\begin{cases} \text{Find } v \in L^2(0, T): \\ \boxed{\int_0^T e^{-\lambda_k t} v(t)\, dt = m_k, \quad k \geq 1.} \end{cases}$$

To solve this problem, we need to answer the following two questions:

1. Does the family $\{e^{-\lambda_k t}\}_{k\geq 1}$ admit a biorthogonal family $\{q_k\}_{k\geq 1}$ in $L^2(0, T)$?
2. If a biorthogonal family $\{q_k\}_{k\geq 1}$ exists, is it possible to estimate $\|q_k\|_{L^2(0,T)}$ as $k \to \infty$?

As a first step, consider $\{e^{-\lambda_k t}\}_{k\geq 1}$ in $L^2(0, \infty)$. Then following Schwartz [18], we have:

Theorem 1.8 *The family* $\{e^{-\lambda_k t}\}_{k\geq 1}$ *is*

1. *complete in* $L^2(0, \infty)$ *if* $\sum_{k\geq 1} 1/\lambda_k = \infty$ *and in this case, it is not minimal;*
2. *minimal in* $L^2(0, \infty)$ *if* $\sum_{k\geq 1} 1/\lambda_k < \infty$ *and in this case, it is not complete.*

Recall that a family $\{x_k\}_{k\geq 1}$ is complete in a Hilbert space H if $\overline{\text{span}\{x_k, k \geq 1\}} = H$; it is minimal if for any $n \geq 1$, $x_n \notin \overline{\text{span}\{x_k, k \geq 1, k \neq n\}}$.

The proof is based on classical properties of the Laplace transform and zeros of holomorphic functions.

Let $f \in L^2(0, \infty)$ and its Laplace transform F given by:

$$F(\lambda) = \int_0^\infty e^{-\lambda t} f(t)\, dt, \quad \Re(\lambda) > 0.$$

The main properties we will use are the following (see for instance [18]):

1. $F \in \mathcal{H}(\mathbb{C}_+)$, the space of holomorphic functions on $\mathbb{C}_+ := \{\lambda \in \mathbb{C}: \Re(\lambda) > 0\}$.

2. For any $\varepsilon > 0$, $F \in \mathcal{H}^\infty(\mathbb{C}_\varepsilon)$ the space of *bounded* holomorphic functions on $\mathbb{C}_\varepsilon = \{\lambda \in \mathbb{C} : \Re(\lambda) > \varepsilon\}$, and moreover

$$\lim_{|\lambda| \to \infty, \lambda \in \mathbb{C}_\varepsilon} F(\lambda) = 0.$$

3. The space $\mathcal{H}^2(\mathbb{C}_+) = \left\{ F \in \mathcal{H}(\mathbb{C}_+) : \int_{\mathbb{R}} |F(\sigma + i\tau)|^2 \, d\tau < \infty, \ \forall \sigma > 0 \right\}$ is a Hilbert space with norm $\|F\|_{\mathcal{H}^2(\mathbb{C}_+)} = \left(\int_{\mathbb{R}} |F(i\tau)|^2 \, d\tau \right)^{1/2}$ and the Laplace transform is an isometry from $L^2(0, \infty)$ in $\mathcal{H}^2(\mathbb{C}_+)$:

$$\|F\|_{\mathcal{H}^2(\mathbb{C}_+)} = \|f\|_{L^2(0,\infty)}, \quad f \in L^2(0, \infty).$$

For bounded holomorphic functions we also have the following properties (see [14]):

Theorem 1.9 *If $f \in \mathcal{H}^\infty(\mathbb{C}_+)$ is a nontrivial function and if $\Lambda = \{z_k\}_{k \geq 1}$ is the sequence of its zeros in $\mathbb{C}_+$, then*

$$\sum_{k \geq 1} \frac{\Re(z_k)}{1 + |z_k|^2} < \infty. \tag{1.16}$$

If (1.16) is satisfied for a sequence $\Lambda = \{z_k\}_{k \geq 1} \subset \mathbb{C}_+$, then the infinite product

$$W(\lambda) = \prod_{k \geq 1} \frac{1 - \lambda/z_k}{1 + \lambda/\overline{z_k}}, \quad \lambda \in \mathbb{C}_+ \tag{1.17}$$

converges absolutely in $\mathbb{C}_+$ and defines a function $W \in \mathcal{H}^\infty(\mathbb{C}_+)$ whose set of zeroes is the sequence Λ.

Proof of Theorem 1.8   Assume that $\sum_{n \geq 1} 1/\lambda_n = \infty$ and let $\varphi \in L^2(0, \infty)$ such that:

$$\forall n \geq 1, \quad \int_0^\infty e^{-\lambda_n t} \varphi(t) \, dt = 0.$$

Let $J : \mathbb{C}_+ \to \mathbb{C}$ be the Laplace transform of φ:

$$J(\lambda) = \int_0^\infty e^{-\lambda t} \varphi(t) \, dt.$$

J is holomorphic on $\mathbb{C}_+$ and uniformly bounded on $\mathbb{C}_\varepsilon = \{\lambda \in \mathbb{C} : \Re(\lambda) > \varepsilon\}$ for all $\varepsilon > 0$. Moreover, by the assumption on φ, $J(\lambda_n) = 0$ for all $n \geq 1$. If φ was nontrivial, from Theorem 1.9 it should follow that

$$\sum_{n \geq 1} \frac{\lambda_n}{1 + \lambda_n^2} < \infty.$$

But this condition is equivalent to $\sum_{n\geq 1} 1/\lambda_n < \infty$ and contradicts the starting assumption: thus $\varphi = 0$.

Therefore, $\{e^{-\lambda_n t}\}_{n\geq 1}$ is complete in $L^2(0,\infty)$ since we have proved that $\text{span}\{e^{-\lambda_n t}, n\geq 1\}^{\perp} = \{0\}$.

However, under the same assumption, it is not minimal since for any $k \geq 1$, the sequence $(\lambda_n)_{\substack{n\geq 1 \\ n\neq k}}$ still has the same properties.

Assume now that $\sum_{n\geq 1} 1/\lambda_n < \infty$. Set

$$W(\lambda) = \prod_{k\geq 1} \frac{1 - \lambda/\lambda_k}{1 + \lambda/\lambda_k}, \quad \lambda \in \mathbb{C}_+. \tag{1.18}$$

If a function J is defined by

$$J(\lambda) = \frac{W(\lambda)}{(1+\lambda)^2}, \quad \lambda \in \mathbb{C}_+ \tag{1.19}$$

then $J \in \mathcal{H}^2(\mathbb{C}_+)$. From the properties of the Laplace transform, there exists a nontrivial function $\varphi \in L^2(0,\infty)$ such that

$$J(\lambda) = \int_0^\infty e^{-\lambda t}\varphi(t)\,dt, \quad \lambda \in \mathbb{C}_+,$$

$$\int_0^\infty |\varphi(t)|^2\,dt = \int_{-\infty}^{+\infty} |J(i\tau)|^2\,d\tau.$$

The λ_n's are the zeros of J and thus it follows that the family $\{e^{-\lambda_n t}\}_{n\geq 1}$ is not complete since φ is nontrivial and belongs to $\overline{\langle e^{-\lambda_n t}, n\geq 1\rangle}^{\perp}$. To prove that $(e^{-\lambda_n t})_{n\geq 1}$ is minimal, a biorthogonal family will be explicitly built for which an estimate of the asymptotic behavior of the norm of its elements will be given. Set for $k \geq 1$,

$$J_k(\lambda) = \frac{J(\lambda)}{J'(\lambda_k)(\lambda - \lambda_k)}, \quad \lambda \in \mathbb{C}_+,$$

where J is the previously defined function. It can easily be proved that $J_k \in \mathcal{H}^2(\mathbb{C}_+)$. Again, from the Laplace transform properties, there exists $\chi_k \in L^2(0,\infty)$ such that:

$$J_k(\lambda) = \int_0^\infty e^{-\lambda t}\chi_k(t)\,dt, \quad \lambda \in \mathbb{C}_+,$$

$$\int_0^\infty |\chi_k(t)|^2\,dt = \int_{-\infty}^{+\infty} |J_k(i\tau)|^2\,d\tau.$$

Since, by definition, $J_k(\lambda_n) = \delta_{kn}$, the biorthogonality of the families $\{e^{-\lambda_n t}\}_{n\geq 1}$ and $\{\chi_n\}_{n\geq 1}$ follows.

To estimate the norm of χ_k, we have from the previous considerations:

$$\int_0^\infty |\chi_k(t)|^2 \, dt = \int_{-\infty}^{+\infty} \left| \frac{J(i\tau)}{J'(\lambda_k)(i\tau - \lambda_k)} \right|^2 \, d\tau, \quad k \geq 1.$$

Since, $|W(i\tau)| = 1$ for all $\tau \in \mathbb{R}$, it appears that:

$$\int_0^\infty |\chi_k(t)|^2 \, dt = \frac{2}{|\lambda_k J'(\lambda_k)|^2} \int_0^\infty \frac{d\tau}{(1 + \tau^2)^2 \left(\left| \frac{\tau}{\lambda_k} \right|^2 + 1 \right)}.$$

By the Lebesgue dominated convergence,

$$\int_0^\infty \frac{d\tau}{(1 + \tau^2)^2 \left(\left| \frac{\tau}{\lambda_k} \right|^2 + 1 \right)} \underset{k \to \infty}{\longrightarrow} \int_0^\infty \frac{d\tau}{(1 + \tau^2)^2} = \frac{\pi}{4}.$$

This leads immediately to:

$$\int_0^\infty |\chi_k(t)|^2 \, dt \underset{k \to \infty}{\sim} \frac{\pi}{2 |\lambda_k J'(\lambda_k)|^2}.$$

The second step in Schwartz's work is the following: consider the closed subspace of $L^2(0, T)$ defined by:

$$A(\Lambda; T) = \overline{\mathrm{Span}\{e^{-\lambda_k t}, k \geq 1\}}^{L^2(0,T)}, \quad 0 < T \leq \infty.$$

Theorem 1.10 *Assume that $\sum_{k \geq 1} 1/\lambda_k < \infty$. The restriction operator*

$$\begin{aligned} R_T: \quad A(\Lambda, \infty) &\to A(\Lambda, T) \\ \varphi &\mapsto \varphi|_{(0,T)} \end{aligned}$$

is an isomorphism. In particular, there exists $C_T > 0$ such that:

$$\|f\|_{L^2(0,\infty)} \leq C_T \|f\|_{L^2(0,T)}, \quad \forall f \in A(\Lambda, \infty).$$

This result is admitted: see [18] for a proof.

With this result in hand, if $e_k(t) = e^{-\lambda_k t}$ for $t \geq 0$ and $k \geq 1$, remark that $R_T e_k = e_k|_{(0,T)}$. Thus,

$$\begin{aligned} \delta_{kj} &= \langle e_k, \chi_j \rangle_{L^2(0,\infty)} \\ &= \left\langle R_T^{-1} R_T \, e_k, \chi_j \right\rangle_{L^2(0,\infty)} \\ &= \left\langle e_k, (R_T^{-1})^* \chi_j \right\rangle_{L^2(0,T)}. \end{aligned}$$

Therefore, the family $\{q_k\}_{k\geq 1} = \{(R_T^{-1})^* \chi_k\}_{k\geq 1}$ is biorthogonal to $\{e^{-\lambda_k t}\}_{k\geq 1}$ in $L^2(0, T)$ and we have the estimate

$$\frac{C_1}{|\lambda_k J'(\lambda_k)|} \leq \|q_k\|_{L^2(0,T)} \leq \frac{C_2}{|\lambda_k J'(\lambda_k)|}, \quad k \geq 1. \qquad (1.20)$$

$\square$

1.5.3 Going Back to the Heat Equation

Anything amounts to estimate:

$$\frac{1}{|\lambda_k J'(\lambda_k)|} = \frac{2(1+\lambda_k)^2}{\prod_{\substack{n\geq 1 \\ n\neq k}} \left|\frac{1-\lambda_k/\lambda_n}{1+\lambda_k/\lambda_n}\right|}$$

if we want to solve problem (1.15). Here, the function J is defined in (1.19). As a consequence of the results of Fattorini and Russell proved in [10, 11], we have in particular that:

Theorem 1.11 *If $\lambda_k = k^2$, then*

$$\lim_{k\to\infty} \frac{\ln\frac{1}{|J'(\lambda_k)|}}{\lambda_k} = 0.$$

In other words, for all $\varepsilon > 0$, there exists $C_\varepsilon > 0$ such that

$$\|q_k\|_{L^2(0,T)} \leq \frac{C}{|\lambda_k J'(\lambda_k)|} \leq C_\varepsilon e^{\varepsilon \lambda_k}, \quad \forall k \geq 1.$$

The control problem (1.14) was reduced to solving the moment problem:

$$\begin{cases} \text{Find } v \in L^2(0, T): \\ f_k \int_0^T e^{-k^2 t} v(t)\, dt = -e^{-k^2 T} y_k^0, \quad k \geq 1. \end{cases}$$

If $f_k \neq 0$ for all k, a formal solution is

$$v = -\sum_{k\geq 1} \frac{e^{-k^2 T}}{f_k} y_k^0 q_k$$

where $\{q_k\}_{k\geq 1} \subset L^2(0, T)$ is the biorthogonal family previously constructed.

The function f can be chosen such that

$$f_k \underset{k\to\infty}{\sim} \frac{C}{k^p} \Rightarrow \forall \varepsilon > 0, \quad \frac{1}{|f_k|} = o(e^{\varepsilon k^2}).$$

But, in view of Theorem 1.11, for any $\varepsilon > 0$:

$$\frac{e^{-k^2 T}}{|f_k|} |y_k^0| \, \|q_k\|_{L^2(0,T)} \leq C_\varepsilon e^{-k^2 T} e^{2\varepsilon k^2} = C_\varepsilon e^{-k^2(T-2\varepsilon)}.$$

Thus $\sum_k e^{-k^2(T-2\varepsilon)} < \infty$ for any $\varepsilon < T/2$. This allows to conclude that

$$v = -\sum_{k \geq 1} \frac{e^{-k^2 T}}{f_k} y_k^0 q_k \in L^2(0, T)$$

and therefore, that the scalar heat equation (1.14) is null-controllable.
Note that the function f could be chosen such that $\text{Supp}(f) \Subset (0, \pi)$.

1.5.4 Example 2: A Minimal Time of Control for a 2×2 Parabolic System due to the Coupling Function

Consider the 2×2 distributed control system:

$$\begin{cases} (\partial_t - \partial_{xx}) y_1 + q(x) y_2 = 0 & \text{in } Q_T, \\ (\partial_t - \partial_{xx}) y_2 = v 1_\omega \\ y(0, \cdot) = 0, \quad y(\pi, \cdot) = 0 & \text{on } (0, T), \\ y(\cdot, 0) = y_0 & \text{in } (0, \pi), \end{cases} \tag{1.21}$$

where

- $q \in L^\infty(0, \pi)$ is a given function, y_0 is the initial datum and $v \in L^2(Q_T)$ is the control function.
- $\omega = (a, b) \subset (0, \pi)$.

The system possesses a unique solution which satisfies

$$y \in L^2(0, T; H_0^1(0, \pi; \mathbb{R}^2)) \cap C^0([0, T]; L^2(0, \pi; \mathbb{R}^2)).$$

Assume that q satisfies

$$\text{supp}(q) \cap \omega = \emptyset \ (\Leftrightarrow \text{supp}(q) \subset [0, a] \cup [b, \pi]). \tag{1.22}$$

For any $k \geq 1$, we associate with the function $q \in L^\infty(0, \pi)$ the sequences $\{I_k(q)\}_{k \geq 1}$ and $\{I_{i,k}(q)\}_{k \geq 1}$, $i = 1, 2$, given by

$$\begin{cases} I_{1,k}(q) := \int_0^a q(x) |\varphi_k(x)|^2 \, dx, \quad I_{2,k}(q) := \int_b^\pi q(x) |\varphi_k(x)|^2 \, dx, \\ I_k(q) := I_{1,k}(q) + I_{2,k}(q) = \int_0^\pi q(x) |\varphi_k(x)|^2 \, dx, \end{cases}$$

where

$$\varphi_k(x) = \sqrt{\frac{2}{\pi}} \sin(kx), \quad \forall x \in (0, \pi), \quad k \geq 1.$$

We will outline the proof of the following result whose details can be found in [5].

Theorem 1.12 *With the previous notations, assume that $q \in L^\infty(0, \pi)$ satisfies (1.22).*

1. *The system is approximately controllable at time $T > 0$ if and only if*

$$|I_k(q)| + |I_{1,k}(q)| \neq 0 \quad \forall k \geq 1.$$

2. *Define*

$$T_0(q) := \varlimsup_{k \to \infty} \frac{min\{-log|I_{1,k}(q)|, -log|I_k(q)|\}}{k^2}.$$

Then:

1. *If $T > T_0(q)$, the system is null-controllable at time T.*
2. *If $T < T_0(q)$, the system is not null-controllable at time T.*

Remark 1.13

- The first point of this theorem has been proved by Boyer and Olive [8]. We will sketch the proof of the second point.
- Note that in [5], the authors show that for any $\delta \in [0, \infty]$, there exists $q \in L^\infty(0, \pi)$ satisfying (1.22) such that $T_0(q) = \delta$. In particular, depending on q, the system may be always approximately controllable and, at the same time, never null-controllable.
- The boundary control problem

$$\begin{cases} (\partial_t - \partial_{xx})y_1 + q(x)y_2 = 0 & \text{in } Q_T, \\ (\partial_t - \partial_{xx})y_2 = 0 & \\ y(0, \cdot) = \begin{pmatrix} 0 \\ 1 \end{pmatrix} v(t), \quad y(\pi, \cdot) = 0 & \text{on } (0, T), \\ y(\cdot, 0) = y_0 & \text{in } (0, \pi), \end{cases}$$

with $v \in L^2(0, T)$, is also studied in [5]. It appears that the minimal time of control is given by:

$$T_b(q) := \varlimsup_{k \to \infty} \frac{-log|I_k(q)|}{k^2}.$$

From this, it follows that boundary and distributed controllability may occur independently.

Set $A_0 = \begin{pmatrix} 0 & 1 \\ 0 & 0 \end{pmatrix}$ and consider the vectorial operator:

$$L := -\frac{d^2}{dx^2} + q(x)A_0 : D(L) \subset L^2(0, \pi; \mathbb{R}^2) \longrightarrow L^2(0, \pi; \mathbb{R}^2)$$

with domain $D(L) = H^2(0, \pi; \mathbb{R}^2) \cap H_0^1(0, \pi; \mathbb{R}^2)$ and also its adjoint L^*. We summarize some properties of the eigenspaces and generalized eigenspaces of these operators in the following proposition:

Proposition 1.14

- The spectra of L and L^* are given by $\sigma(L) = \sigma(L^*) = \{k^2 : k \geq 1\}$.
- Given $k \geq 1$, let ψ_k be the unique solution of the nonhomogeneous Sturm–Liouville problem:

$$\begin{cases} -\psi_{xx} - k^2\psi = [I_k(q) - q(x)]\varphi_k, & \text{in } (0, \pi), \\ \psi(0) = 0, \quad \psi(\pi) = 0, \\ \displaystyle\int_0^\pi \psi(x)\varphi_k(x)\, dx = 0. \end{cases}$$

- The family $\mathcal{B} = \left\{ \Phi_{1,k} = \begin{pmatrix} \varphi_k \\ 0 \end{pmatrix}, \quad \Phi_{2,k} = \begin{pmatrix} \psi_k \\ \varphi_k \end{pmatrix} \right\}_{k \geq 1}$ satisfies

$$(L - k^2 I_d)\Phi_{1,k} = 0 \quad \text{and} \quad (L - k^2 I_d)\Phi_{2,k} = I_k(q)\Phi_{1,k}.$$

- The family $\mathcal{B}^* = \left\{ \Phi_{1,k}^* := \begin{pmatrix} \varphi_k \\ \psi_k \end{pmatrix}, \quad \Phi_{2,k}^* := \begin{pmatrix} 0 \\ \varphi_k \end{pmatrix} \right\}_{k \geq 1}$ is biorthogonal to $\mathcal{B}$ and

$$\left(L^* - k^2 I_d\right)\Phi_{1,k}^* = I_k(q)\Phi_{2,k}^* \quad \text{and} \quad \left(L^* - k^2 I_d\right)\Phi_{2,k}^* = 0.$$

- In particular, if $I_k \neq 0$ then k^2 is a simple eigenvalue and $\Phi_{1,k}$ and $\Phi_{2,k}$ (resp., $\Phi_{2,k}^*$ and $\Phi_{1,k}^*$) are, respectively, an eigenfunction and a generalized eigenfunction of the operator L (resp., L^*) associated with k^2, while if $I_k = 0$ then $\Phi_{1,k}$ and $\Phi_{2,k}$ are both eigenfunctions of L (resp., L^*) associated with k^2.
- $\mathcal{B}$ and $\mathcal{B}^*$ are Riesz bases in $L^2(0, \pi; \mathbb{R}^2)$ and for any $y^0 \in L^2(0, \pi; \mathbb{R}^2)$

$$\begin{aligned} y^0 &= \sum_{k \geq 1} \left\{ \langle y^0, \Phi_{1,k}^* \rangle \Phi_{1,k} + \langle y^0, \Phi_{2,k}^* \rangle \Phi_{2,k} \right\} \\ &= \sum_{k \geq 1} \left\{ y_{1,k}^0 \Phi_{1,k} + y_{2,k}^0 \Phi_{2,k} \right\}. \end{aligned}$$

If we look for the solution of System (1.21) in the form:

$$y(t) = \sum_{k \geq 1} \{y_{1,k}(t)\Phi_{1,k} + y_{2,k}(t)\Phi_{2,k}\}$$

we readily get the system that $\{y_{i,k}, i = 1, 2; k \geq 1\}$ must solve the sequence of 2×2 differential systems:

$$\begin{cases} y'_{1,k} + k^2 y_{1,k} + I_k(q)y_{2,k} = \left\langle Bv1_\omega, \Phi^*_{1,k} \right\rangle \\ y'_{2,k} + k^2 y_{2,k} = \left\langle Bv1_\omega, \Phi^*_{2,k} \right\rangle \\ (y_{1,k}, y_{2,k})|_{t=0} = \left(y^0_{1,k}, y^0_{2,k} \right) \end{cases}$$

with $B = \begin{pmatrix} 0 \\ 1 \end{pmatrix}$.

Solving this system, we get by setting $v_{i,k}(t) = \left\langle Bv1_\omega, \Phi^*_{i,k} \right\rangle$ for $i = 1, 2$:

$$y_{1,k}(T) = e^{-k^2 T}\left(y^0_{1,k} - TI_k y^0_{2,k} \right)$$
$$+ \int_0^T e^{-k^2(T-t)}[v_{1,k}(t) - (T-t)I_k v_{2,k}(t)]dt,$$

$$y_{2,k}(T) = e^{-k^2 T}y^0_{2,k} + \int_0^T e^{-k^2(T-t)}v_{2,k}(t)\, dt.$$

Then, $y(T) = 0$ if and only if

$$\begin{cases} \int_0^T e^{-k^2(T-t)}[v_{1,k}(t) - (T-t)I_k v_{2,k}(t)]dt = -e^{-k^2 T}\left(y^0_{1,k} + TI_k y^0_{2,k} \right), \\ \int_0^T e^{-k^2(T-t)}v_{2,k}(t)\, dt = -e^{-k^2 T}y^0_{2,k}. \end{cases}$$

If we look for v in the form:

$$v(x, t) = f_1(x)v_1(T - t) + f_2(x)v_2(T - t), \quad (x, t) \in Q_T,$$

where $v_1, v_2 \in L^2(0, T)$ are new controls, only depending on t, and $f_1, f_2 \in L^2(0, \pi)$ are appropriate functions satisfying the condition $\mathrm{supp}(f_1)$, $\mathrm{supp}(f_2) \subseteq \omega = (a, b)$, then we get the system:

$$\begin{cases} f_{1,k}\displaystyle\int_0^T v_1(t)e^{-k^2 t}\, dt + f_{2,k}\int_0^T v_2(t)e^{-k^2 t}\, dt = -e^{-k^2 T}\left\langle y_0, \Phi^*_{2,k} \right\rangle \\ \tilde{f}_{1,k}\displaystyle\int_0^T v_1(t)e^{-k^2 t}\, dt + \tilde{f}_{2,k}\int_0^T v_2(t)e^{-k^2 t}\, dt \\ \qquad - I_k(q)f_{1,k}\int_0^T v_1(t)te^{-k^2 t}\, dt - I_k(q)f_{2,k}\int_0^T v_2(t)te^{-k^2 t}\, dt \\ \qquad = -e^{-k^2 T}\left(\left\langle y_0, \Phi^*_{1,k} \right\rangle - TI_k(q)\left\langle y_0, \Phi^*_{2,k} \right\rangle \right), \end{cases}$$

$$(1.23)$$

where, for $k \geq 1, \widetilde{f}_{1,k}, \widetilde{f}_{2,k}$ are given by

$$\widetilde{f}_{i,k} := \int_0^\pi f_i(x)\psi_k(x)\,dx, \quad f_{i,k} := \int_0^\pi f_i(x)\varphi_k(x)\,dx, \quad i = 1, 2.$$

Remark that, if we fix $k \geq 1$, it is a linear system of two equations and four unknown quantities:

$$\int_0^T v_i(t)e^{-k^2 t}\,dt, \quad \int_0^T v_i(t)te^{-k^2 t}\,dt, \quad i = 1, 2.$$

Working on system (1.23), it is possible to prove the following result:

Lemma 1.15 The moment problem (1.23) has the form

$$\begin{cases} \displaystyle\int_0^T v_i(t)e^{-k^2 t}\,dt = e^{-k^2 T} M_{1,i}^{(k)}(y_0), \\[2mm] \displaystyle\int_0^T v_i(t)te^{-k^2 t}\,dt = e^{-k^2 T} M_{2,i}^{(k)}(y_0), \end{cases}$$

where the quantities $M_{i,j}^{(k)}(y_0) \in \mathbb{R}$, with $k \geq 1$ and $1 \leq i, j \leq 2$, satisfy the following property: for any $\varepsilon > 0$ there exists a positive constant C_ε (only depending on ε) such that

$$\left| M_{i,j}^{(k)}(y_0) \right| \leq C_\varepsilon e^{k^2(T_0(q)+2\varepsilon)} \|y_0\|_{L^2(0,\pi;\mathbb{R}^2)}, \quad \forall k \geq 1, \quad 1 \leq i, j \leq 2.$$

The conclusion giving the positive null controllability result is based on the following two observations (proved in [3]):

- The family $\{e^{-k^2 t}, te^{-k^2 t}\}_{k \geq 1}$ is minimal in $L^2(0, T)$.
- There exists a biorthogonal family $\{q_{1,k}, q_{2,k}\}_{k \geq 1}$ in $L^2(0, T)$ such that

$$\|q_{i,k}\|_{L^2(0,T)}^2 \leq C_\varepsilon e^{\varepsilon k^2}, \quad i = 1, 2; \ k \geq 1. \tag{1.24}$$

The formal solution of the moment problem is then given by

$$v_i(t) = \sum_{k \geq 1} \left\{ e^{-k^2 T} M_{1,i}^{(k)}(y_0)q_{1,k} + e^{-k^2 T} M_{2,i}^{(k)}(y_0) \right\}, \quad i = 1, 2.$$

With the previous estimates in Lemma 1.15 and (1.24), it can be checked exactly as in Section 1.5.3, that $v_i \in L^2(0, T)$ for $T > T_0(q)$ and leads to the first point of the theorem.

Let now assume that $T < T_0(q)$ and consider the adjoint problem:

$$\begin{cases} -\theta_t - \theta_{xx} + q(x)A_0^*\theta = 0 & \text{in } Q_T, \\ \theta(0, \cdot) = 0, \quad \theta(\pi, \cdot) = 0 & \text{on } (0, T), \\ \theta(\cdot, T) = \theta^0 & \text{in } (0, \pi). \end{cases}$$

The idea, is to find a sequence of initial data $\left(\theta_k^0\right)_{k\geq 1}$ for which

$$\frac{\iint_{\omega\times(0,T)}|B^*\theta_k(x,t)|^2\,dx\,dt}{\|\theta_k(\cdot,0)\|_{L^2(0,\pi;\mathbb{R}^2)}^2}\underset{k\to\infty}{\to}0$$

where θ_k is the solution of the adjoint problem associated with θ_k^0.

In this way, the observability inequality

$$\|\theta(\cdot,0)\|_{L^2(0,\pi;\mathbb{R}^2)}^2\leq C\int_{\omega\times(0,T)}|B^*\theta(x,t)|^2\,dx\,dt$$

fails.

For $\theta_k^0=a_k\Phi_{1,k}^*+b_k\Phi_{2,k}^*$, with $(a_k,b_k)\in\mathbb{R}^2$, the solution of the adjoint problem is given by:

$$\theta_k(t,x)=a_k e^{-k^2(T-t)}\left(\Phi_{1,k}^*-(T-t)I_k(q)\Phi_{2,k}^*\right)+b_k e^{-k^2(T-t)}\Phi_{2,k}^*.$$

Computing, we get

$$\|\theta_k(\cdot,0)\|_{L^2(0,\pi;\mathbb{R}^2)}^2$$
$$=e^{-2k^2T}\left\{|a_k|^2+\left[|a_k|^2\|\psi_k\|_{L^2(0,\pi)}^2+(b_k-Ta_kI_k(q))^2\right]\right\}$$
$$\geq e^{-2k^2T}|a_k|^2$$

and

$$\int_0^T\!\!\int_\omega|B^*\theta(x,t)|^2=\int_0^T\!\!\int_\omega e^{-2k^2t}\left|a_k\psi_k(x)+(b_k-ta_kI_k(q))\varphi_k(x)\right|^2\,dx.$$

It can be proved that for any $x\in\omega$

$$\psi_k(x)=\tau_k\varphi_k(x)-I_k(q)g_k(x)-\sqrt{\frac{\pi}{2}}\frac{1}{k}I_{1,k}(q)\cos(kx).$$

Thus:

$$\int_0^T\!\!\int_\omega|B^*\theta(x,t)|^2=\int_0^T\!\!\int_\omega e^{-2k^2t}\Big|(a_k\tau_k+b_k)\varphi_k(x)$$
$$-a_kI_k(q)(g_k(x)+t\varphi_k(x))-\sqrt{\frac{\pi}{2}}\frac{a_k}{k}I_{1,k}(q)\cos(kx)\Big|^2\,dx\,dt.$$

Choosing $a_k=1$ and $b_k=-\tau_k$ gives the inequality:

$$\int_0^T\!\!\int_\omega|B^*\theta(x,t)|^2\leq C\left(|I_{1,k}(q)|^2+|I_k(q)|^2\right)$$

and to summarize:

$$e^{-k^2 T} \le C\left(|I_{1,k}(q)|^2 + |I_k(q)|^2\right)$$
$$\le C e^{-2k^2\left[\frac{1}{k^2}\min\left(-\log|I_{1,k}(q)|, -\log I_k(q)|\right)\right]}.$$

The contradiction follows with a suitable choice of a subsequence $\{k_n\}$ in connection with the definition of $T_0(q)$.

1.5.5 Example 3: A Minimal Time of Control Due to the Condensation of the Eigenvalues of the System

Consider the system

$$
\begin{cases}
y_1' = \dfrac{\partial^2 y_1}{\partial x^2} & Q_T = (0, T) \times (0, \pi) \\[2mm]
y_2' = d\dfrac{\partial^2 y_2}{\partial x^2} & Q_T \\[2mm]
y_i(t, 0) = b_i v(t), \quad y_i(t, \pi) = 0 & i = 1, 2 \\[2mm]
y_i(0, \cdot) = y_i^0 & i = 1, 2
\end{cases}
\tag{1.25}
$$

where $0 < d < 1$, $b_i \in \mathbb{R}$ $(i = 1, 2)$ and $v \in L^2(0, T)$.
For $y^0 = (y_1^0, y_2^0) \in H^{-1}(0, \pi; \mathbb{R}^2)$, there exists a unique solution

$$y \in C\left([0, T]; H^{-1}(0, \pi; \mathbb{R}^2)\right) \cap L^2(Q_T).$$

Indeed, the solution is given by:

$$y_i = \sum_{k \ge 1} y_{i,k}\varphi_k$$

where $\varphi_k(x) = \sqrt{\frac{2}{\pi}}\sin(kx)$ and

$$y_{1,k}(t) = e^{-k^2 t}y_{1,k}^0 + \sqrt{\frac{2}{\pi}}kb_1 \int_0^t e^{-k^2(t-s)}v(s)\,ds$$

$$y_{2,k}(t) = e^{-dk^2 t}y_{2,k}^0 + \sqrt{\frac{2}{\pi}}kb_2 \int_0^t e^{-dk^2(t-s)}v(s)\,ds.$$

Thus the null controllability issue reduces to: find $v \in L^2(0, T)$ such that

$$
\begin{cases}
\sqrt{\frac{2}{\pi}} k b_1 \int_0^T e^{-k^2 t} v(t - t)\, ds = -e^{-k^2 T} y_{1,k}^0 \\[2mm]
\sqrt{\frac{2}{\pi}} k b_2 \int_0^T e^{-dk^2 t} v(t - t)\, ds = -e^{-dk^2 T} y_{2,k}^0
\end{cases}
, \quad k \geq 1. \qquad (1.26)
$$

Exercise 1.1

1. A first necessary condition for solvability of (1.26) for any initial data is

$$ b_i \neq 0, \quad i = 1, 2. $$

2. A second necessary condition for solvability of (1.26) for any initial data is

$$ dk^2 \neq \ell^2, \quad \forall k, \ell \geq 1. $$

(This last condition is equivalent to $\sqrt{d} \notin \mathbb{Q}$.)

With these two necessary conditions, we now have to solve:

$$
\begin{cases}
\int_0^T e^{-k^2 t} v(t - t)\, ds = -e^{-k^2 T} \dfrac{y_{1,k}^0}{\sqrt{\frac{2}{\pi}} k b_1} \\[4mm]
\int_0^T e^{-dk^2 t} v(t - t)\, ds = -e^{-dk^2 T} \dfrac{y_{2,k}^0}{\sqrt{\frac{2}{\pi}} k b_2}
\end{cases}
, \quad k \geq 1.
$$

So, this time, we are dealing with the family $\{e^{-k^2 t}, e^{-dk^2 t}\}$. Set

$$ \lambda_{2k} = dk^2, \quad \lambda_{2k+1} = k^2, \quad k \geq 1. $$

Then clearly $\sum_{n \geq 1} 1/\lambda_n < \infty$ and it follows from Theorem 1.8 that $\{e^{-\lambda_n t}\}$ is minimal in $L^2(0, T)$ and a biorthogonal family $\{q_n\}$ can be found such that

$$
\frac{C_1}{|\lambda_n J'(\lambda_n)|} \leq \|q_n\|_{L^2(0,T)}^2 \leq \frac{C_2}{|\lambda_n J'(\lambda_n)|}, \quad n \geq 1 \qquad (1.27)
$$

with, let us recall (see (1.19)):

$$
\frac{1}{|\lambda_n J'(\lambda_n)|} = \frac{2(1 + \lambda_n)^2}{\prod\limits_{\substack{\ell \geq 1 \\ n \neq \ell}} \left| \frac{1 - \lambda_n / \lambda_\ell}{1 + \lambda_n / \lambda_\ell} \right|}.
$$

Again, a formal solution is given by:

$$
v(t - t) = \sum_{n \geq 1} e^{-\lambda_n T} m_n q_n \qquad (1.28)
$$

and as in the Fattorini–Russell example (Sections 1.5.1 and 1.5.3), we need an estimate of $1/|\lambda_n J'(\lambda_n)|$ or, in other words, to compute

$$\varlimsup_{k \to \infty} \log \frac{1}{|\lambda_n J'(\lambda_n)|}.$$

Remember that if $\lambda_n = n^2$, we had

$$\varlimsup_{k \to \infty} \log \frac{1}{|n^2 J'(n^2)|} = 0.$$

Indeed, it has been proven (see [17] for instance) that for any ordered real sequence

$$|\lambda_n - \lambda_m| \geq \alpha\,|n - m| \implies \varlimsup_{k \to \infty} \log \frac{1}{|\lambda_n J'(\lambda_n)|} = 0.$$

But for the sequence $\{\lambda_{2k} = dk^2,\ \lambda_{2k+1} = k^2\}$, there does not exist a positive real number such that $\sqrt{d} \notin \mathbb{Q}$ and for which this separability condition satisfied. Indeed, it is proven in [4] that:

Proposition 1.16 For any $c \in [0, \infty]$, there exists $\sqrt{d} \notin \mathbb{Q}$ such that for $\{\lambda_{2k} = dk^2,\ \lambda_{2k+1} = k^2\}$

$$\varlimsup_{k \to \infty} \frac{\log \frac{1}{|\lambda_n J'(\lambda_n)|}}{\lambda_n} = c.$$

Actually, the number

$$C(\Lambda) := \varlimsup_{k \to \infty} \frac{\log \frac{1}{|\lambda_n J'(\lambda_n)|}}{\lambda_n} \tag{1.29}$$

associated with the sequence $\Lambda = \{\lambda_n\}$ has a name: it is the **index of condensation** of the sequence $\{\lambda_n\}$.

Before saying more about this index, let us see to what kind of conclusion leads the introduction of this number:

Theorem 1.17 *Assume that*

$$b_i \neq 0 \ (i = 1, 2) \quad and \quad \sqrt{d} \notin \mathbb{Q}.$$

Let $c(\Lambda)$ be the condensation index associated with $\Lambda = \{\lambda_{2k} = dk^2,\ \lambda_{2k+1} = k^2\}$.
Then:

1. *If $T > c$, then the system is null-controllable.*
2. *If $T < c$, then the system is not null-controllable.*

To prove the first point, it suffices to use the definition of $C(\Lambda)$ in (1.29) and the estimate (1.27 to prove that the function v defined in 1.28 belongs to $L^2(0, T)$. The second point is more tricky: we need some of the intermediate results given in the forthcoming section.

1.6 The Index of Condensation

1.6.1 Definition

Definition 1.18 Let $\Lambda = (\lambda_k)_{k\geq 1}$ be an increasing sequence of real numbers. A condensation grouping of Λ is any sequence of sets $G=(G_k)_{k\geq 1}$ satisfying the following properties:

1. $\Lambda \cap G_k \neq \varnothing$ for all $k \geq 1$ and $\Lambda = \cup_{k\geq 1}(\Lambda \cap G_k)$.
2. If

$$\Lambda \cap G_k = \{\lambda_{n_k}, \lambda_{n_k+1}, \ldots, \lambda_{n_k+p_k}\}, \quad k \geq 1$$

then

$$\lim_{k\to\infty} \frac{p_k}{\lambda_{n_k}} = 0$$

$$\lim_{k\to\infty} \frac{\lambda_{n_k+p_k}}{\lambda_{n_k}} = 1.$$

The second item of this definition characterizes what is meant by *condensation*.

Let $G=(G_k)_{k\geq 1}$ a condensation grouping of Λ.

- For all $k \geq 1$, the index of condensation of $G_k = \{\lambda_{n_k}, \ldots, \lambda_{n_k+p_k}\}$ is the number

$$\delta(G_k) = \sup_{0\leq l\leq p_k} \frac{1}{|\lambda_{n_k+l}|} \ln\left(\frac{p_k!}{\prod_{\substack{0\leq j\leq p_k \\ j\neq l}} |(\lambda_{n_k+l} - \lambda_{n_k+j})|}\right).$$

- The index of condensation of $G=(G_k)_{k\geq 1}$ is the number: $\delta(G) = \overline{\lim}_{k\to\infty} \delta(G_k)$.

Definition 1.19 The index of condensation of $\Lambda = (\lambda_k)_{k\geq 1}$ is the number $\delta(\Lambda)$ defined to be the supremum of the set $\{\delta(G)\}$ where G may be any condensation grouping.

Example 1.20 Let $\Lambda = (\lambda_n)$ and set $G_n = \{\lambda_n\}$.
  $G = (G_n)$ is a condensation grouping with $p_n = 0$ and

$$\delta(G_n) = 0, \quad n \geq 1.$$

Conclusion: $\boxed{\delta(\Lambda) \geq 0 \text{ for any } \Lambda.}$

Example 1.21 Let $\alpha \geq 1$ and $\beta > 0$ and set $\Lambda = \{\lambda_{2n} = n^\alpha, \ \lambda_{2n+1} = n^\alpha + e^{-n^\beta},$
$n \geq 1\}$.
  Define: $G_n = \,]\lambda_{2n} - \frac{1}{2}; \lambda_{2n} + \frac{1}{2}[, \quad n \geq 1.$ $G = (G_n)$ is a condensation grouping with $p_n = 1$ and

$$\delta(G_n) = \max\left\{ \frac{1}{n^\alpha} \ln \frac{1!}{e^{-n^\beta}}, \ \frac{1}{n^\alpha + e^{-n^\beta}} \ln \frac{1!}{e^{-n^\beta}} \right\}$$

$$= \frac{1}{n^\alpha} \ln \frac{1}{e^{-n^\beta}} = n^{\beta - \alpha}.$$

Thus:

$$\boxed{\delta(G) = \limsup_{n \to \infty} \delta(G_n) = \begin{cases} 0, & 0 < \beta < \alpha, \\ 1, & \beta = \alpha, \\ \infty, & \beta > \alpha. \end{cases}}$$

1.6.2 Optimal Condensation Grouping

Definition 1.22 Let $\Lambda = (\lambda_n)_{n \geq 1}$ be a real increasing sequence. A sequence Λ is **measurable** if there exists $D \in [0, \infty[$ such that $\lim_{n \to \infty} \dfrac{n}{\lambda_n} = D$. The number D when it exists is the density of Λ.

Example 1.23 Let $\alpha \geq 1$ and $\beta > 0$ and set $\Lambda = \{\lambda_{2n} = n^\alpha, \ \lambda_{2n+1} = n^\alpha + e^{-n^\beta},$
$n \geq 1\}$. Then

$$D = \begin{cases} 0, & \text{if } \alpha > 1 \\ 1, & \text{if } \alpha = 1 \\ \infty, & \text{if } \alpha < 1. \end{cases}$$

  For any subset $E \subset \mathbb{R}$, $\mathcal{N}_\Lambda(E) = \mathrm{card}(\Lambda \cap E)$ is the counting function of E. The following intermediate result is due to Shackell [17]:

Lemma 1.24 Let $\Lambda = (\lambda_k)_{k \geq 1}$ be an increasing sequence of real numbers, with finite density D, and let $0 < q < 1/(2D + \eta)$ where $\eta > 0$ is some arbitrary fixed number. For all $\lambda \in \Lambda$ and any integer $r \geq 1$, let $I(\lambda, rq) = \,]\lambda - rq, \lambda + rq[.$

Then there exists a greatest integer $p(\lambda)$ such that

$$\mathcal{N}_\Lambda(I(\lambda, p(\lambda)\, q)) \geq p(\lambda).$$

With the previous lemma in hand, we have the following result:

Theorem 1.25 [17] *Let* $\Lambda = (\lambda_n)_{n \geq 1}$ *be an increasing sequence of positive real numbers, with density* D *and* $0 < q < 1/(2D + \eta)$ *where* $\eta > 0$ *is a fixed arbitrary number.*

Then there exists a condensation grouping $G = (G_k)_{k \geq 1}$ *such that:*

1. $\delta(G) = \delta(\Lambda)$.
2. *Fix* $k \geq 1$. *For all* $\lambda \in \Lambda \cap G_k$ *and* $(\mu_n)_{1 \leq n \leq m} \subset \Lambda \backslash (\Lambda \cap G_k)$, *we have*

$$\prod_{n=1}^m |\lambda - \mu_n| \geq q^m m!.$$

Proof [**Sketch of the proof**] By successive applications of the previous Lemma, define a sequence of intervals in the following way. Let:

$$G_1 = I(\lambda_1, p_1 q), \quad n_1 = 1, \quad p_1 = p(\lambda_{n_1}).$$

Denote by λ_{n_2} the smallest element of Λ not belonging to $I_1(q)$ and set

$$G_2 = I(\lambda_{n_2}, p_2 q), \qquad p_2 = p(\lambda_{n_2}).$$

Let $k \geq 1$ and suppose constructed the intervals $(G_j)_{1 \leq j \leq k}$. Denote by $\lambda_{n_{k+1}}$ the smallest element of Λ not belonging to $\cup_{1 \leq j \leq k} G_j$. We then define

$$G_{k+1} = I(\lambda_{n_{k+1}}, p_{k+1} q), \qquad p_{k+1} = p(\lambda_{n_{k+1}}).$$

The sequence thus defined satisfies the conclusion of the theorem as it can be easily checked. □

Example 1.26 Let $\alpha \geq 1$ and $\beta > 0$ and set $\Lambda = \{\lambda_{2n} = n^\alpha, \; \lambda_{2n+1} = n^\alpha + e^{-n^\beta}, \; n \geq 1\}$.

Define: $G_n = \,]\lambda_{2n} - \tfrac{1}{2}; \lambda_{2n} + \tfrac{1}{2}[, \; n \geq 1$.

$G = (G_n)$ is a condensation grouping with $p_n = 1$ and

$$\delta(G_n) = \max\left\{ \frac{1}{n^\alpha} \ln \frac{1!}{e^{-n^\beta}}, \; \frac{1}{n^\alpha + e^{-n^\beta}} \ln \frac{1!}{e^{-n^\beta}} \right\}$$

$$= \frac{1}{n^\alpha} \ln \frac{1}{e^{-n^\beta}} = n^{\beta - \alpha}.$$

Thus:

$$\delta(G) = \limsup_{n\to\infty} \delta(G_n) = \left\{ \begin{array}{ll} 0, & 0 < \beta < \alpha, \\ 1, & \beta = \alpha, \\ \infty, & \beta > \alpha. \end{array} \right.$$

With this previous construction, it can be checked that G is optimal and thus: $\delta(G) = \delta(\Lambda)$.

1.6.3 Interpolating Function

To the sequence $\Lambda = (\lambda_n)_{n\geq1}$ with density $D \geq 0$, is associated the interpolating function

$$C(\lambda) = \prod_{n\geq1} \left(1 - \frac{\lambda^2}{\lambda_n^2} \right), \quad \lambda \in \mathbb{C}.$$

The following result is due to Bernstein [7] for real sequences and Shackell [17] for complex sequences.

Theorem 1.27 *Let $\Lambda = (\lambda_n)_{n\geq1}$ be an increasing real sequence of positive numbers, measurable with finite density $D \geq 0$. Then its index of condensation $\delta(\Lambda)$ is given by:*

$$\delta(\Lambda) = \overline{\lim_{k\to\infty}} \frac{\ln \frac{1}{|C'(\lambda_k)|}}{\lambda_k}.$$

The property that links with the boundary control problem (1.25) of the previous section is the following:

Theorem 1.28 *If $\Lambda = (\lambda_n)_{n\geq1}$ is an increasing sequence of positive numbers such that $\sum_{n\geq1} 1/\lambda_n < \infty$, then*

$$\delta(\Lambda) = \overline{\lim_{k\to\infty}} \frac{\ln \frac{1}{|J'(\lambda_k)|}}{\lambda_k}$$

where

$$|J'(\lambda_k)| = \frac{\prod_{\substack{n\geq1 \\ n\neq k}} \left| \frac{1 - \lambda_k/\lambda_n}{1 + \lambda_k/\lambda_n} \right|}{2\lambda_k(1 + \lambda_k)^2}.$$

Remark 1.29 Note that the condensation index is defined even for sequences which do not satisfy the condition $\sum_{n\geq1} 1/\lambda_n < \infty$.

Exercise 1.2 Define $\Lambda = (\lambda_n)_{n \geq 1}$ by

$$\lambda_{k^2+l} = k^2 + le^{-k^\beta}, \quad k \geq 1, \quad 0 \leq l \leq 2k, \quad (\beta > 0)$$

and prove that:

$$\delta(\Lambda) = \begin{cases} \infty, & \beta > 1 \\ 2, & \beta = 1 \\ 0, & 0 < \beta < 1 \end{cases}.$$

1.6.4 An Interpolating Formula of Jensen

An interpolation formula due to Jensen [15] assures that if f is a holomorphic function on a convex domain $\Omega \subset \mathbb{C}$ and $A = \{a_j\}_{0 \leq j \leq q} \subset \Omega$ is a set of distinct points, then there exists $\theta \in [-1, 1]$ and $\xi \in \mathrm{Conv}(A)$, the convex hull of A, such that

$$\sum_{j=0}^{q} \frac{f(a_j)}{P'_A(a_j)} = \frac{\theta}{q!} \frac{d^q f}{dz^q}(\xi),$$

where for any finite set F the function P_F is defined by:

$$P_F(\lambda) = \prod_{\mu \in F} (\lambda - \mu).$$

As a consequence of this formula, we have:

Theorem 1.30 *Let $\Lambda = \{\lambda_k\}_{k \geq 1}$ be an increasing sequence of real numbers and $G = \{G_k\}_{k \geq 1}$ any condensation grouping associated with Λ. Then:*

$$\lim_{k \to \infty} \int_0^\infty \left| \sum_{\lambda_n \in G_k} \frac{p_k!}{P'_{G_k}(\lambda_n)} e^{-\lambda_n t} \right|^2 dt = 0.$$

This result will prove decisive to establish noncontrollability results.

1.6.5 Going Back to the Boundary Control Problem

The associated observability inequality with the boundary control problem (1.25) is of the form:

$$\int_0^T \left| \sum_{k \geq 1} c_k e^{-\lambda_k t} \right|^2 dt \geq C_T \sum_{k \geq 1} e^{-\lambda_k T} c_k^2, \quad \forall c \in \ell^2.$$

Let $G = (G_k)$ be the optimal condensation grouping given by Theorem 1.25. Fix $k \geq 1$ and set

$$c_n^k = \begin{cases} \dfrac{p_k!}{P'_{G_k}(\lambda_n)}, & \text{if } \lambda_n \in G_k, \\ 0 & \text{otherwise.} \end{cases} \tag{1.30}$$

Then, from Theorem 1.30:

$$\int_0^T \left| \sum_{n \geq 1} c_n^k e^{-\lambda_n t} \right|^2 dt = \int_0^T \left| \sum_{\lambda_n \in G_k} \frac{p_k!}{P'_{G_k}(\lambda_n)} e^{-\lambda_n t} \right|^2 dt \underset{k \to \infty}{\to} 0.$$

On the other hand, if $T < \delta(\Lambda)$

$$\sum_{n \geq 1} \left| c_n^k \right|^2 e^{-\lambda_n T} = \sum_{\lambda_n \in G_n} \left| \frac{p_k!}{P'_{G_k}(\lambda_n)} e^{-\lambda_n T} \right|^2$$

$$\geq \left| e^{\lambda_{n_k}(\delta(\Lambda) - \varepsilon - T)} \right|^2 \to \infty.$$

This implies that the observability inequality does not hold and concludes the proof of the Theorem 1.17.

References

[1] F. Alabau-Boussouira, M. Léautaud, Indirect controllability of locally coupled wave-type systems and applications, *J. Math. Pures Appl.* (9) **99** (2013), no. 5, 544–76.

[2] F. Ammar Khodja, A. Benabdallah, C. Dupaix, Null-controllability of some reaction-diffusion systems with one control force, *J. Math. Anal. Appl.* **320** (2006), no. 2, 928–43.

[3] F. Ammar-Khodja, A. Benabdallah, M. González-Burgos, L. de Teresa, The Kalman condition for the boundary controllability of coupled parabolic systems. Bounds on biorthogonal families to complex matrix exponentials, *J. Math. Pures Appl.* **96** (2011), no. 6, 555–90.

[4] F. Ammar Khodja, A. Benabdallah, M. Gonzalez-Burgos, L. de Teresa. Minimal time for the null controllability of parabolic systems: The effect of the condensation index of complex sequences. *J. Funct. Anal.* **267** (2014), no. 7, 2077–151.

[5] F. Ammar Khodja, A. Benabdallah, M. González-Burgos, L. de Teresa, New phenomena for the null controllability of parabolic systems: Minimal time and geometrical dependence. https://hal.archives-ouvertes.fr/hal-01165713 (2015).

[6] C. Bardos, G. Lebeau, J. Rauch, Sharp sufficient conditions for the observation, control, and stabilization of waves from the boundary, *SIAM J. Control Optim.* **30** (1992), no. 5, 1024–65.

[7]   V. Bernstein, *Leçons sur les Progrès Récents de la Théorie des Séries de Dirichlet*, Gauthier-Villars, Paris, 1933.

[8]   F. Boyer, G. Olive, Approximate controllability conditions for some linear 1D parabolic systems with space-dependent coefficients, *Math. Control Relat. Fields* **4** (2014), no. 3, 263–87.

[9]   J.-M. Coron. *Control and Nonlinearity*, Mathematical Surveys and Monographs, 136, American Mathematical Society, Providence, RI, 2007.

[10] H.O. Fattorini, D. L. Russell, Exact controllability theorems for linear parabolic equations in one space dimension, *Arch. Rational Mech. Anal.* **43** (1971), 272–92.

[11] H.O. Fattorini, D. L. Russell, Uniform bounds on biorthogonal functions for real exponentials with an application to the control theory of parabolic equations, *Quart. Appl. Math.* **32** (1974/5), 45–69.

[12] A. V. Fursikov, O. Yu. Imanuvilov, *Controllability of Evolution Equations*, Lecture Notes Series, 34. Seoul National University, Research Institute of Mathematics, Global Analysis Research Center, Seoul, 1996.

[13] M. González-Burgos, L. de Teresa, Controllability results for cascade systems of *m* coupled parabolic PDEs by one control force, *Port. Math.* **67** (2010), no. 1, 91–113.

[14] E. Hille. *Analytic Function Theory*. Vol. II. Second edition. AMS Chelsea Publishing, Boston, MA, 2000.

[15] J.L.W.V. Jensen, Sur une expression simple du reste dans la formule d'interpolation de Newton, *Bull. Acad. Roy. Danemark* **1894** (1894), 246–52.

[16] G. Lebeau, L. Robbiano, Contrôle exact de l'équation de la chaleur, *Comm. Partial Diff. Eq.* **20** (1995), no. 1–2, 335–56.

[17] J. R. Shackell, Overconvergence of Dirichlet series with complex exponents, *J. Analyse Math.* **22** (1969), 135–70.

[18] L. Schwartz. *Étude des Sommes d'Exponentielles Réelles*. Hermann, Paris, France, 1943.

LABORATOIRE DE MATHÉMATIQUES DE BESANÇON
E-mail address: farid.ammar-khodja@univ-fcomte.fr

2

Stabilization of Semilinear PDEs, and Uniform Decay under Discretization

EMMANUEL TRÉLAT

Abstract

These notes are issued from a short course given by the author in a summer school in Chambéry in June 2015.

We consider general semilinear PDEs and we address the following two questions:

1. How to design an efficient feedback control locally stabilizing the equation asymptotically to 0?
2. How to construct such a stabilizing feedback from approximation schemes?

To address these issues, we distinguish between parabolic and hyperbolic semilinear PDEs. By parabolic, we mean that the linear operator underlying the system generates an analytic semigroup. By hyperbolic, we mean that this operator is skew-adjoint.

We first recall general results allowing one to consider the nonlinear term as a perturbation that can be absorbed when one is able to construct a Lyapunov function for the linear part. We recall in particular some known results borrowed from the Riccati theory.

However, since the numerical implementation of Riccati operators is computationally demanding, we focus then on the question of being able to design "simple" feedbacks. For parabolic equations, we describe a method consisting of designing a stabilizing feedback, based on a small finite-dimensional (spectral) approximation of the whole system. For hyperbolic equations, we focus on simple linear or nonlinear feedbacks and we investigate the question of obtaining sharp decay results.

When considering discretization schemes, the decay obtained in the continuous model cannot in general be preserved for the discrete model, and we address the question of adding appropriate viscosity terms in the numerical scheme, in order to recover a uniform decay. We consider space, time, and then full discretizations and we report in particular on the most recent results obtained in the literature.

Finally, we describe several open problems and issues.

Mathematics Subject Classification 2010. 35B37,74B05,93B05

Key words and phrases. Null controllability, thermoelastic plate system, single distributed control

Contents

2.1 Introduction and General Results

2.1.1 General Setting

Let X and U be Hilbert spaces. We consider the semilinear control system

$$\dot{y}(t) = Ay(t) + F(y(t)) + Bu(t), \qquad (2.1)$$

where $A : D(A) \to X$ is an operator on X, generating a C_0 semigroup, $B \in L(U, D(A^*)')$, and $F : X \to X$ is a (nonlinear) mapping of class C^1, assumed to be Lipschitz on the bounded sets of X, with $F(0) = 0$. We refer to [10, 18, 43, 44] for well-posedness of such systems (existence, uniqueness, appropriate functional spaces, etc.).

We focus on the following two questions:

1. How to design an efficient feedback control $u = Ky$, with $K \in L(X, U)$, locally stabilizing (2.1) asymptotically to 0?
2. How to construct such a stabilizing feedback from approximation schemes?

Moreover, we want the feedback to be as simple as possible, in order to promote a simple implementation. Given the fact that the decay obtained in the continuous setting may not be preserved under discretization, we are also interested in the way one should design a numerical scheme, in order to get a uniform decay for the solutions of the approximate system, i.e., in order to guarantee uniform properties with respect to the discretization parameters $\triangle x$ and/or $\triangle t$. This is the objective of these notes, to address those issues.

Concerning the first point, note that, in general, stabilization cannot be global because there may exist other steady states than 0, i.e., $\bar{y} \in X$ such that $A\bar{y} + F(\bar{y}) = 0$. This is why we speak of local stabilization. Without loss of generality we focus on the steady-state 0 (otherwise, just design a feedback of the kind $u = K(y - \bar{y})$).

We are first going to recall well-known results on how to obtain local stabilization results, first in finite dimension, and then in infinite dimension. As a preliminary remark, we note that, replacing if necessary A with $A + dF(0)$, we can always assume, without loss of generality, that $dF(0) = 0$, and thus,

$$\|F(y)\| = o(\|y\|) \quad \text{near } y = 0.$$

Then, a first possibility to stabilize (2.1) locally at 0 is to consider the nonlinear term $F(y)$ as a perturbation, that we are going to absorb with a linear feedback. Let us now recall standard results and methods.

2.1.2 In Finite Dimension

In this section, we assume that $X = \mathbb{R}^n$. In that case, A is a square matrix of size n, and B is a matrix of size $n \times m$. Then, as it is well known, stabilization is doable under the Kalman condition

$$\text{rank}(B, AB, \ldots, A^{n-1}B) = n,$$

and there are several ways to do it (see, e.g., [42] for a reference on what follows).

2.1.2.1 First Possible Way: By Pole-Shifting

According to the *pole-shifting theorem*, there exists K (matrix of size $m \times n$) such that $A + BK$ is Hurwitz, that is, all its eigenvalues have negative real part. Besides, according to the *Lyapunov lemma*, there exists a positive definite symmetric matrix P of size n, such that

$$P(A + BK) + (A + BK)^\top P = -I_n.$$

It easily follows that the function V defined on $\mathbb{R}^n$ by

$$V(y) = y^\top P y$$

is a Lyapunov function for the closed-loop system $\dot{y}(t) = (A + BK)y(t)$.

Now, for the semilinear system (2.1) in closed loop with $u(t) = Ky(t)$, we have

$$\frac{d}{dt}V(y(t)) = -\|y(t)\|^2 + y(t)^\top PF(y(t)) \leq -C_1\|y(t)\|^2 \leq -C_2 V(y(t)),$$

under an a priori assumption on $y(t)$, for some positive constants C_1 and C_2, whence the local asymptotic stability.

2.1.2.2 Second Possible Way: By Riccati Procedure

The Riccati procedure consists of computing the unique negative definite solution of the *algebraic Riccati equation*

$$A^\top E + EA + EBB^\top E = I_n.$$

Then, we set $u(t) = B^\top Ey(t)$.

Note that this is the control minimizing $\int_0^{+\infty}(\|y(t)\|^2 + \|u(t)\|^2)dt$ for the control system $\dot{y}(t) = Ay(t) + Bu(t)$. This is one of the best well-known results of the linear quadratic theory in optimal control.

Then the function

$$V(y) = -y^* Ey$$

is a Lyapunov function, as before.

Now, for the semilinear system (2.1) in closed loop with $u(t) = B^\top Ey(t)$, we have

$$\frac{d}{dt}V(y(t)) = -y(t)^\top(I_n + EBB^\top E)y(t) - y(t)^\top EF(y(t))$$
$$\leq -C_1\|y(t)\|^2 \leq -C_2 V(y(t)),$$

under an a priori assumption on $y(t)$, and we easily infer the local asymptotic stability property.

2.1.3 In Infinite Dimension

2.1.3.1 Several Reminders

When the Hilbert space X is infinite-dimensional, several difficulties occur with respect to the finite-dimensional setting. To explain them, we consider the uncontrolled linear system

$$\dot{y}(t) = Ay(t), \tag{2.2}$$

with $A: D(A) \to X$ generating a C_0 semigroup $S(t)$.

The first difficulty is that none of the following three properties are equivalent:

1. $S(t)$ is exponentially stable, i.e., there exist $C > 0$ and $\delta > 0$ such that $\|S(t)\| \leq Ce^{-\delta t}$, for every $t \geq 0$.
2. The spectral abscissa is negative, i.e., $\sup\{\text{Re}(\lambda) \mid \lambda \in \sigma(A)\} < 0$.
3. All solutions of (2.2) converge to 0 in X, i.e., $S(t)y^0 \xrightarrow[t \to +\infty]{} 0$, for every $y^0 \in X$.

For example, if we consider the linear wave equation with local damping

$$y_{tt} - \triangle y + \chi_\omega y = 0,$$

in some domain Ω of $\mathbb{R}^n$, with Dirichlet boundary conditions, and with ω an open subset of Ω, then it is always true that any solution (y, y_t) converges to $(0, 0)$ in $H_0^1(\Omega) \times L^2(\Omega)$ (see [16]). Besides, it is known that we have exponential stability if and only if ω satisfies the geometric control condition (GCC). This condition says, roughly, that any generalized ray of geometric optics must meet ω in finite time. Hence, if for instance Ω is a square, and ω is a small ball in Ω, then GCC does not hold and hence the exponential stability fails, whereas convergence of solutions to the equilibrium is valid.

In general, we always have

$$\sup\{\text{Re}(\lambda) \mid \lambda \in \sigma(A)\} \leq \inf\{\mu \in \mathbb{R} \mid \exists C > 0, \, \|S(t)\| \leq Ce^{\mu t} \quad \forall t \geq 0\},$$

in other words the spectral abscissa is always less than or equal to the best exponential decay rate. The inequality may be strict, and the equality is referred to as "spectral growth condition."

Let us go ahead by recalling the following known results:

- *Datko theorem*: $S(t)$ is exponentially stable if and only if, for every $y^0 \in X$, $S(t)y^0$ converges exponentially to 0.
- *Arendt–Batty theorem*: If there exists $M > 0$ such that $\|S(t)\| \leq M$ for every $t \geq 0$, and if $i\mathbb{R} \subset \rho(A)$ (where $\rho(A)$ is the resolvent set of A), then $S(t)y^0 \xrightarrow[t \to +\infty]{} 0$ for every $y^0 \in X$.
- *Huang–Prüss theorem*: Assume that there exists $M > 0$ such that $\|S(t)\| \leq M$ for every $t \geq 0$. Then $S(t)$ is exponentially stable if and only if $i\mathbb{R} \subset \rho(A)$ and $\sup_{\beta \in \mathbb{R}} \|(i\beta \text{id} - A)^{-1}\| < +\infty$.

Finally, we recall that:

- Exactly null-controllable implies exponentially stabilizable, meaning that there exists $K \in L(X, U)$ such that $A + BK$ generates an exponentially stable C_0 semigroup.
- Approximately controllable does not imply exponentially stabilizable.

For all reminders done here, we refer to [15, 18, 31, 32, 45].

2.1.3.2 Stabilization

Let us now consider the linear control system

$$\dot{y} = Ay + Bu$$

and let us first assume that $B \in L(U, X)$, that is, the control operator B is bounded. We assume that the pair (A, B) is exponentially stabilizable.

Riccati procedure. As before, the Riccati procedure consists of finding the unique negative definite solution $E \in L(X)$ of the *algebraic Riccati equation*

$$A^* E + EA + EBB^* E = \mathrm{id},$$

in the sense of $\langle (2EA + EBB^* E - \mathrm{id})y, y \rangle = 0$, for every $y \in D(A)$, and then of setting $u(t) = B^* Ey(t)$. Note that this is the control minimizing $\int_0^{+\infty} (\|y(t)\|^2 + \|u(t)\|^2)\, dt$ for the control system $\dot{y} = Ay + Bu$.

Then, as before, the function $V(y) = -\langle y, Ey \rangle$ is a Lyapunov function for the system in closed loop $\dot{y} = (A + BK)y$.

Now, for the semilinear system (2.1) in closed loop with $u(t) = B^* Ey(t)$, we have

$$\frac{d}{dt} V(y(t)) = -\langle y(t), (\mathrm{id} + EBB^* E)y(t) \rangle - \langle y(t), EF(y(t)) \rangle$$

$$\leq -C_1 \|y(t)\|^2 \leq -C_2 V(y(t))$$

and we infer local asymptotic stability.

For $B \in L(U, D(A^*)')$ unbounded, things are more complicated. Roughly, the theory is complete in the parabolic case (i.e., when A generates an analytic semigroup), but is incomplete in the hyperbolic case (see [29, 30] for details).

Rapid stabilization. An alternative method exists in the case where A generates a *group* $S(t)$, and $B \in L(U, D(A^*)')$ an admissible control operator (see [44] for the notion of admissibility). An example covered by this setting is the wave equation with Dirichlet control.

The strategy developed in [27] consists of setting

$$u = -B^* C_\lambda^{-1} y \qquad \text{with} \qquad C_\lambda = \int_0^{T+1/2\lambda} f_\lambda(t) S(-t) BB^* S(-t)^* \, dt,$$

with $\lambda > 0$ arbitrary, $f_\lambda(t) = e^{-2\lambda t}$ if $t \in [0, T]$ and $f_\lambda(t) = 2\lambda e^{-2\lambda T}(T + 1/2\lambda - t)$ if $t \in [T, T + 1/2\lambda]$. Besides, the function

$$V(y) = \langle y, C_\lambda^{-1} y \rangle$$

is a Lyapunov function (as noticed in [11]). Actually, this feedback yields exponential stability with rate $-\lambda$ for the closed-loop system $\dot{y} = (A - BB^* C_\lambda^{-1})y$, whence the wording "rapid stabilization" since $\lambda > 0$ is arbitrary.

Then, thanks to that Lyapunov function, the above rapid stabilization procedure applies as well to the semilinear control system (2.1), yielding a local stabilization result (with any exponential rate).

2.1.4 Existing Results for Discretizations

We recall hereafter existing convergence results for space semidiscretizations of the Riccati procedure. We denote by E_N the approximate Riccati solution, expecting that $E_N \to E$ as $N \to +\infty$.

One can find in [7, 8, 22, 26, 32] a general result showing convergence of the approximations E_N of the Riccati operator E, under assumptions of *uniform* exponential stabilizability, and of *uniform* boundedness of E_N:

$$\|S_{A_N + B_N B_N^\top E_N}(t)\| \leq C e^{-\delta t}, \qquad \|E_N\| \leq M,$$

for every $t \geq 0$ and every $N \in \mathbb{N}^*$, with uniform positive constants C, δ and M.

In [7, 29, 32], the convergence of E_N to E is proved in the general parabolic case, for unbounded control operators, that is, when $A : D(A) \to X$ generates an analytic semigroup, $B \in L(U, D(A^*)')$, and (A, B) is exponentially stabilizable.

The situation is therefore definitive in the parabolic setting. In contrast, if the semigroup $S(t)$ is not analytic, the theory is not complete. Uniform exponential stability is proved under uniform Huang-Prüss conditions in [32]. More precisely, it is proved that, given a sequence $(S_n(\cdot))$ of C_0 semigroups on X_n, of generators A_n, $(S_n(\cdot))$ is uniformly exponentially stable if and only if $i\mathbb{R} \subset \rho(A_n)$ for every $n \in \mathbb{N}$ and

$$\sup_{\beta \in \mathbb{R}, n \in \mathbb{N}} \|(i\beta \mathrm{id} - A_n)^{-1}\| < +\infty.$$

This result is used, e.g., in [37], to prove uniform stability of second-order equations with (bounded) damping and with viscosity term, under uniform gap condition on the eigenvalues.

This result also allows to obtain convergence of the Riccati operators, for second-order systems

$$\ddot{y} + Ay = Bu,$$

with $A : D(A) \to X$ positive self-adjoint, with compact inverse, and $B \in L(U, X)$ (bounded control operator).

But the approximation theory for general LQR problems remains incomplete in the general hyperbolic case with unbounded control operators, for instance it is not done for wave equations with Dirichlet boundary control.

2.1.5 Conclusion

Concerning implementation issues, solving Riccati equations (or computing Gramians, in the case of rapid stabilization) in large dimension is computationally demanding. In what follows, we would like to find other ways to proceed.

Our objective is therefore to design *simple* feedbacks with *efficient* approximation procedures.

In the sequel, we are going to investigate two situations:

1. *Parabolic* case (A generates an analytic semigroup): we are going to show how to design feedbacks based on a small number of spectral modes.
2. *Hyperbolic* case, i.e., $A = -A^*$ in (2.1): we are going to consider two "simple" feedbacks:
   - Linear feedback $u = -B^* y$: in that case, if $F = 0$ then $\frac{1}{2}\frac{d}{dt}\|y(t)\|^2 = -\|B^* y(t)\|^2$. We will investigate the question of how to ensure uniform exponential decay for approximations.
   - Nonlinear feedback $u = B^* G(y)$: we will also investigate the question of how to ensure uniform (sharp) decay for approximations.

2.2 Parabolic PDEs

In this section, we assume that the operator A in (2.1) generates an analytic semigroup. Our objective is to design stabilizing feedbacks based on a small number of spectral modes.

To simplify the exposition, we consider a 1D semilinear heat equation, and we will comment further on extensions.

Let $L > 0$ be fixed and let $f: \mathbb{R} \to \mathbb{R}$ be a function of class C^2 such that $f(0) = 0$. Following [12], We consider the 1D semilinear heat equation

$$y_t = y_{xx} + f(y), \qquad y(t, 0) = 0, \quad y(t, L) = u(t), \qquad (2.3)$$

where the state is $y(t, \cdot): [0, L] \to \mathbb{R}$ and the control is $u(t) \in \mathbb{R}$.

We want to design a feedback control locally stabilizing (2.3) asymptotically to 0. Note that this cannot be global, because we can have other steady states (a steady state is a function $y \in C^2(0, L)$ such that $y''(x) + f(y(x)) = 0$ on $(0, L)$ and $y(0) = 0$). By the way, here, without loss of generality we consider the steady-state 0.

Let us first note that, for every $T > 0$, (2.3) is well posed in the Banach space

$$Y_T = L^2(0, T; H^2(0, L)) \cap H^1(0, T; L^2(0, L)),$$

which is continuously embedded in $L^\infty((0, T) \times (0, L))$.[1]

First of all, in order to end up with a Dirichlet problem, we set

$$z(t, x) = y(t, x) - \frac{x}{L} u(t).$$

Assuming (for the moment) that u is differentiable, we set $v(t) = u'(t)$, and we consider in the sequel v as a control. We also assume that $u(0) = 0$. Then we have

$$z_t = z_{xx} + f'(0)z + \frac{x}{L}f'(0)u - \frac{x}{L}v + r(t, x), \qquad z(t, 0) = z(t, L) = 0,$$
$$(2.4)$$

with $z(0, x) = y(0, x)$ and

$$r(t, x) = \left(z(t, x) + \frac{x}{L}u(t)\right)^2 \int_0^1 (1 - s)f''\left(sz(s, x) + s\frac{x}{L}u(s)\right) ds.$$

Note that, given $B > 0$ arbitrary, there exist positive constants C_1 and C_2 such that, if $|u(t)| \le B$ and $\|z(t, \cdot)\|_{L^\infty(0,L)} \le B$, then

$$\|r(t, \cdot)\|_{L^\infty(0,L)} \le C_1(u(t)^2 + \|z(t, \cdot)\|^2_{L^\infty(0,L)}) \le C_2(u(t)^2 + \|z(t, \cdot)\|^2_{H^1_0(0,L)}).$$

---

[1] Indeed, considering $v \in L^2(0, T; H^2(0, L))$ with $v_t \in H^1(0, T; L^2(0, L))$, writing $v = \sum_{j,k} c_{jk} e^{ijt} e^{ikx}$, we have

$$\sum_{j,k} |c_{jk}| \le \left(\sum_{j,k} \frac{1}{1 + j^2 + k^4}\right)^{1/2} \left(\sum_{j,k}(1 + j^2 + k^4)|c_{jk}|^2\right)^{1/2},$$

and these series converge, whence the embedding, allowing to give a sense to $f(y)$. Now, if y_1 and y_2 are solutions of (2.3) on $[0, T]$, then $y_1 = y_2$. Indeed, $v = y_1 - y_2$ is a solution of $v_t = v_{xx} + av$, $v(t, L) = v(t, L) = 0$, $v(0, x) = 0$, with $a(t, x) = g(y_1(t, x), y_2(t, x))$ where g is a function of class C^1. We infer that $v = 0$.

In the sequel, $r(t, x)$ will be considered as a remainder.

We define the operator $A = \triangle + f'(0)\mathrm{id}$ on $D(A) = H^2(0, L) \cap H_0^1(0, L)$, so that (2.4) is written as

$$\dot{u} = v, \qquad z_t = Az + au + bv + r, \qquad z(t, 0) = z(t, L) = 0, \qquad (2.5)$$

with $a(x) = \frac{x}{L} f'(0)$ and $b(x) = -\frac{x}{L}$.

Since A is self-adjoint and has a compact resolvent, there exists a Hilbert basis $(e_j)_{j \geq 1}$ of $L^2(0, L)$, consisting of eigenfunctions $e_j \in H_0^1(0, L) \cap C^2([0, L])$ of A, associated with eigenvalues $(\lambda_j)_{j \geq 1}$ such that $-\infty < \cdots < \lambda_n < \cdots < \lambda_1$ and $\lambda_n \to -\infty$ as $n \to +\infty$.

Any solution $z(t, \cdot) \in H^2(0, L) \cap H_0^1(0, L)$ of (2.4), as long as it is well defined, can be expanded as a series

$$z(t, \cdot) = \sum_{j=1}^{\infty} z_j(t) e_j(\cdot)$$

(converging in $H_0^1(0, L)$), and then we have, for every $j \geq 1$,

$$\dot{z}_j(t) = \lambda_j z_j(t) + a_j u(t) + b_j v(t) + r_j(t),$$

with

$$a_j = \frac{f'(0)}{L} \int_0^L x e_j(x)\, dx, \quad b_j = -\frac{1}{L} \int_0^L x e_j(x)\, dx, \quad r_j(t) = \int_0^L r(t, x) e_j(x)\, dx.$$

Setting, for every $n \in \mathbb{N}^*$,

$$X_n(t) = \begin{pmatrix} u(t) \\ z_1(t) \\ \vdots \\ z_n(t) \end{pmatrix}, \ A_n = \begin{pmatrix} 0 & 0 & \cdots & 0 \\ a_1 & \lambda_1 & \cdots & 0 \\ \vdots & \vdots & \ddots & \vdots \\ a_n & 0 & \cdots & \lambda_n \end{pmatrix}, \ B_n = \begin{pmatrix} 1 \\ b_1 \\ \vdots \\ b_n \end{pmatrix}, \ R_n(t) = \begin{pmatrix} 0 \\ r_1(t) \\ \vdots \\ r_n(t) \end{pmatrix},$$

we have, then,

$$\dot{X}_n(t) = A_n X_n(t) + B_n v(t) + R_n(t).$$

Lemma 2.1 The pair (A_n, B_n) satisfies the Kalman condition.

Proof We compute

$$\det(B_n, A_n B_n, \ldots, A_n^n B_n) = \prod_{j=1}^{n} (a_j + \lambda_j b_j) \mathrm{VdM}(\lambda_1, \ldots, \lambda_n), \qquad (2.6)$$

where $\mathrm{VdM}(\lambda_1, \ldots, \lambda_n)$ is a Van der Monde determinant, and thus is never equal to zero since the λ_i, $i = 1, \ldots, n$, are pairwise distinct. On the other part, using the fact that each e_j is an eigenfunction of A and belongs to $H_0^1(0, L)$, we compute

$$a_j + \lambda_j b_j = \frac{1}{L} \int_0^L x(f'(0) - \lambda_j)e_j(x)\, dx = -\frac{1}{L} \int_0^L x e_j''(x)\, dx = -e_j'(L),$$

and this quantity is never equal to zero since $e_j(L) = 0$ and e_j is a nontrivial solution of a linear second-order scalar differential equation. Therefore the determinant (2.6) is never equal to zero. □

By the pole-shifting theorem, there exists $K_n = (k_0, \dots, k_n)$ such that the matrix $A_n + B_n K_n$ has -1 as an eigenvalue of multiplicity $n + 1$. Moreover, by the Lyapunov lemma, there exists a symmetric positive definite matrix P_n of size $n + 1$ such that

$$P_n(A_n + B_n K_n) + (A_n + B_n K_n)^\top P_n = -I_{n+1}.$$

Therefore, the function defined by $V_n(X) = X^\top P_n X$ for any $X \in \mathbb{R}^{n+1}$ is a Lyapunov function for the closed-loop system $\dot{X}_n(t) = (A_n + B_n K_n)X_n(t)$.

Let $\gamma > 0$ and $n \in \mathbb{N}^*$ to be chosen later. For every $u \in \mathbb{R}$ and every $z \in H^2(0, L) \cap H_0^1(0, L)$, we set

$$V(u, z) = \gamma X_n^\top P_n X_n - \frac{1}{2} \langle z, Az \rangle_{L^2(0,L)} = \gamma X_n^\top P_n X_n - \frac{1}{2} \sum_{j=1}^\infty \lambda_j z_j^2, \quad (2.7)$$

where $X_n \in \mathbb{R}^{n+1}$ is defined by $X_n = (u, z_1, \dots, z_n)^\top$ and $z_j = \langle z(\cdot), e_i(\cdot) \rangle_{L^2(0,L)}$ for every j.

Using that $\lambda_n \to -\infty$ as $n \to +\infty$, it is clear that, choosing $\gamma > 0$ and $n \in \mathbb{N}^*$ large enough, we have $V(u, z) > 0$ for all $(u, z) \in \mathbb{R} \times (H^2(0, L) \cap H_0^1(0, L)) \setminus \{(0, 0)\}$. More precisely, there exist positive constants C_3, C_4, C_5 and C_6 such that

$$C_3 \left(u^2 + \|z\|_{H_0^1(0,L)}^2 \right) \leq V(u, z) \leq C_4 \left(u^2 + \|z\|_{H_0^1(0,L)}^2 \right),$$

$$V(u, z) \leq C_5 \left(\|X_n\|_2^2 + \|Az\|_{L^2(0,L)}^2 \right), \qquad \gamma C_6 \|X_n\|_2^2 \leq V(u, z),$$

for all $(u, z) \in \mathbb{R} \times (H^2(0, L) \cap H_0^1(0, L))$. Here, $\| \ \|_2$ designates the Euclidean norm of $\mathbb{R}^{n+1}$.

Our objective is now to prove that V is a Lyapunov function for the system (2.5) in closed loop with the control $v = K_n X_n$.

In what follows, we thus take $v = K_n X_n$ and u defined by $\dot{u} = v$ and $u(0) = 0$. We compute

$$\frac{d}{dt} V(u(t), z(t)) = -\gamma \|X_n(t)\|_2^2 - \|Az(t, \cdot)\|_{L^2}^2 - \langle Az(t, \cdot), a(\cdot)\rangle_{L^2} u(t)$$
$$-\langle Az(t, \cdot), b(\cdot)\rangle_{L^2} K_n X_n(t) - \langle Az(t, \cdot), r(t, \cdot)\rangle_{L^2}$$
$$+\gamma\left(R_n(t)^\top P_n X_n(t) + X_n(t)^\top P_n R_n(t)\right). \qquad (2.8)$$

Let us estimate the terms at the right-hand side of (2.8). Under the a priori estimates $|u(t)| \leq B$ and $\|z(t, \cdot)\|_{L^\infty(0,L)} \leq B$, there exist positive constants C_7, C_8 and C_9 such that

$$|\langle Az, a\rangle_{L^2} u| + |\langle Az, b\rangle_{L^2} K_n X_n| \leq \frac{1}{4}\|Az\|_{L^2}^2 + C_7\|X_n\|_2^2,$$

$$|\langle Az, r\rangle_{L^2}| \leq \frac{1}{4}\|Az\|_{L^2}^2 + C_8 V^2, \qquad \|R_n\|_\infty \leq \frac{C_2}{C_3} V,$$

$$|\gamma\left(R_n^\top P_n X_n + X_n^\top P_n R_n\right)| \leq \frac{C_2}{C_3\sqrt{C_6}}\sqrt{\gamma}\, V^{3/2}.$$

We infer that, if $\gamma > 0$ is large enough, then there exist positive constants C_{10} and C_{11} such that $\frac{d}{dt} V \leq -C_{10} V + C_{11} V^{3/2}$. We easily conclude the local asymptotic stability of the system (2.5) in closed loop with the control $v = K_n X_n$.

Remark 2.2 Of course, the above local asymptotic stability may be achieved with other procedures, for instance, by using the Riccati theory. However, the procedure developed here is much more efficient because it consists of stabilizing a finite-dimensional part of the system, mainly, the part that is not naturally stable. We refer to [12] for examples and for more details. Actually, we have proved in that reference that, thanks to such a strategy, we can pass from any steady-state to any other one, provided that the two steady states belong to a same connected component of the set of steady states: this is a partially global exact controllability result.

The main idea used above is the following fact, already used in the remarkable early paper [38]. Considering the linearized system with no control, we have an infinite-dimensional linear system that can be aligned, through a spectral decomposition, in two parts: the first part is finite-dimensional, and consists of all spectral modes that are unstable (meaning that the corresponding eigenvalues have nonnegative real part); the second part is infinite-dimensional, and consists of all spectral modes that are asymptotically stable (meaning that the corresponding eigenvalues have negative real part). The idea used here then consists of focusing on the finite-dimensional unstable part of the system, and to design a feedback control in order to stabilize that part. Then, we plug this control in the infinite-dimensional system, and we have to check that this feedback indeed stabilizes the whole system (in the sense that it does not destabilize the other infinite-dimensional part). This is the role of the Lyapunov function V defined by (2.7).

The extension to general systems (2.1) is quite immediate, at least in the parabolic setting under appropriate spectral assumptions (see [39] for Couette flows and [14] for Navier–Stokes equations).

But it is interesting to note that it does not work only for parabolic equations: this idea has been as well used in [13] for the 1D semilinear equation

$$y_{tt} = y_{xx} + f(y), \quad y(t,0) = 0, \ y_x(t,L) = u(t),$$

with the same assumptions on f as before. We first note that, if $f(y) = cy$ is linear (with $c \in L^\infty(0,L)$), then, setting $u(t) = -\alpha y_t(t,L)$ with $\alpha > 0$ yields an exponentially decrease of the energy $\int_0^L (y_t(t,x)^2 + y_x(t,x)^2) \, dt$, and moreover, the eigenvalues of the corresponding operator have a real part tending to $-\infty$ as α tends to 1. Therefore, in the general case, if α is sufficiently close to 1 then at most a finite number of eigenvalues may have a nonnegative real part. Using a Riesz spectral expansion, the same kind of method as the one developed above can therefore be applied, and yields a feedback based on a finite number of modes, that stabilizes locally the semilinear wave equation, asymptotically to equilibrium.

2.3 Hyperbolic PDEs

In this section, we assume that the operator A in (2.1) is skew-adjoint, that is,

$$A^* = -A, \quad D(A^*) = D(A).$$

Let us start with a simple remark. If $F = 0$ (linear case), then, choosing the very simple linear feedback $u = -B^* y$ and setting $V(y) = \frac{1}{2} \frac{d}{dt} \|y\|_X^2$, we have

$$\frac{d}{dt} V(y(t)) = -\|B^* y(t)\|_X^2 \le 0,$$

and then we expect that, under reasonable assumptions, we have exponential asymptotic stability (and this will be the case under observability assumptions, as we are going to see).

Now, if we choose a nonlinear feedback $u = B^* G(y)$, we ask the same question: what are sufficient conditions ensuring asymptotic stability, and if so, with which sharp decay?

Besides, we will investigate the following important question: how to ensure *uniform* properties when discretizing?

2.3.1 The Continuous Setting

2.3.1.1 Linear Case

In this section, we assume that $F = 0$ (linear case), and we assume that B is bounded. Taking the linear feedback $u = -B^* y$ as said above, we have the closed-loop system $\dot{y} = Ay - BB^* y$. For convenience, in what follows we rather write this equation in the form (more standard in the literature)

$$\dot{y}(t) + Ay(t) + By(t) = 0, \tag{2.9}$$

where A is a densely defined skew-adjoint operator on X and B is a bounded nonnegative self-adjoint operator on X (we have just replaced A with $-A$ and BB^* with B).

We start hereafter with the question of the exponential stability of solutions of (2.9).

Equivalence between Observability and Exponential Stability. The following result is a generalization of the main result of [23].

Theorem 2.3 *Let X be a Hilbert space, let $A : D(A) \to X$ be a densely defined skew-adjoint operator, let B be a bounded self-adjoint nonnegative operator on X. We have equivalence of:*

1. *There exist $T > 0$ and $C > 0$ such that every solution of the conservative equation*

$$\dot{\phi}(t) + A\phi(t) = 0 \tag{2.10}$$

   satisfies the observability inequality

$$\|\phi(0)\|_X^2 \leq C \int_0^{T_0} \|B^{1/2}\phi(t)\|_X^2 \, dt.$$

2. *There exist $C_1 > 0$ and $\delta > 0$ such that every solution of the damped equation*

$$\dot{y}(t) + Ay(t) + By(t) = 0 \tag{2.11}$$

satisfies

$$E_y(t) \le C_1 E_y(0) e^{-\delta t},$$

where $E_y(t) = \frac{1}{2} \|y(t)\|_X^2$.

Proof Let us first prove that the first property implies the second one: we want to prove that every solution of (2.11) satisfies

$$E_y(t) = \frac{1}{2} \|y(t)\|_X^2 \le E_y(0) e^{-\delta t} = \frac{1}{2} \|y(0)\|_X^2 e^{-\delta t}.$$

Consider ϕ solution of (2.10) with $\phi(0) = y(0)$. Setting $\theta = y - \phi$, we have

$$\dot\theta + A\theta + By = 0, \quad \theta(0) = 0.$$

Then, taking the scalar product with θ, since A is skew-adjoint, we get $\langle \dot\theta + By, \theta \rangle_X = 0$. But, setting $E_\theta(t) = \frac{1}{2} \|\theta(t)\|_X^2$, we have $\dot E_\theta = -\langle By, \theta \rangle_X$. Then, integrating a first time over $[0, t]$, and then a second time over $[0, T]$, since $E_\theta(0) = 0$, we get

$$\int_0^T E_\theta(t)\, dt = -\int_0^T \int_0^t \langle By(s), \theta(s) \rangle_X \, ds\, dt$$

$$= -\int_0^T (T - t)\langle B^{1/2} y(t), B^{1/2}\theta(t) \rangle_X \, dt,$$

where we have used the Fubini theorem. Hence, thanks to the Young inequality $ab \le \frac{\alpha}{2} a^2 + \frac{1}{2\alpha} b^2$ with $\alpha = 2$, we infer that

$$\frac{1}{2} \int_0^T \|\theta(t)\|_X^2 \, dt \le T \|B^{1/2}\| \int_0^T \|B^{1/2} y(t)\|_X \|\theta(t)\|_X \, dt$$

$$\le T^2 \|B^{1/2}\|^2 \int_0^T \|B^{1/2} y(t)\|_X^2 \, dt + \frac{1}{4} \int_0^T \|\theta(t)\|_X^2 \, dt,$$

and therefore,

$$\int_0^T \|\theta(t)\|_X^2 \, dt \le 4T^2 \|B^{1/2}\|_X^2 \int_0^T \|B^{1/2} y(t)\|_X^2 \, dt.$$

Now, since $\phi = y - \theta$, it follows that

$$\int_0^T \|B^{1/2}\phi(t)\|_X^2 \, dt \le 2 \int_0^T \|B^{1/2} y(t)\|_X^2 \, dt + 2 \int_0^T \|B^{1/2}\theta(t)\|_X^2 \, dt$$

$$\le (2 + 8T^2 \|B^{1/2}\|^4) \int_0^T \|B^{1/2} y(t)\|_X^2 \, dt.$$

Finally, since

$$E_y(0) = E_\phi(0) = \frac{1}{2}\|\phi(0)\|_X^2 \leq \frac{C}{2} \int_0^T \|B^{1/2}\phi(t)\|_X^2 \, dt$$

it follows that

$$E_y(0) \leq C(1 + 4T^2\|B^{1/2}\|^4) \int_0^T \|B^{1/2}y(t)\|_X^2 \, dt.$$

Besides, one has $E_y'(t) = -\|B^{1/2}y(t)\|_X^2$, and then $\int_0^T \|B^{1/2}y(t)\|_X^2 \, dt = E_y(0) - E_y(T)$. Therefore

$$E_y(0) \leq C(1 + 4T^2\|B^{1/2}\|^4)(E_y(0) - E_y(T)) = C_1(E_y(0) - E_y(T))$$

and hence

$$E_y(T) \leq \frac{C_1 - 1}{C_1} E_y(0) = C_2 E_y(0),$$

with $C_2 < 1$.

Actually this can be done on every interval $[kT, (k+1)T]$, and it yields $E_y((k+1)T) \leq C_2 E_y(kT)$ for every $k \in \mathbb{N}$, and hence $E_y(kT) \leq E_y(0)C_2^k$.

For every $t \in [kT, (k+1)T)$, noting that $k = [\frac{t}{T}] > \frac{t}{T} - 1$, and that $\ln \frac{1}{C_2} > 0$, it follows that

$$C_2^k = \exp(k \ln C_2) = \exp(-k \ln \frac{1}{C_2}) \leq \frac{1}{C_2} \exp\left(\frac{-\ln \frac{1}{C_2}}{T}t\right)$$

and hence $E_y(t) \leq E_y(kT) \leq \delta E_y(0)\exp(-\delta t)$ for some $\delta > 0$.

Let us now prove the converse: assume the exponential decrease of solutions of (2.11), and let us prove the observability property for solutions of (2.10).

From the exponential decrease inequality, one has

$$\int_0^T \|B^{1/2}y(t)\|_X^2 \, dt = E_y(0) - E_y(T) \geq (1 - C_1 e^{-\delta T})E_y(0) = C_2 E_y(0),$$

$$(2.12)$$

and for $T > 0$ large enough there holds $C_2 = 1 - C_1 e^{-\delta T} > 0$.

Then we make the same proof as before, starting from (2.10), that we write in the form

$$\dot{\phi} + A\phi + B\phi = B\phi,$$

and considering the solution of (2.11) with $y(0) = \phi(0)$. Setting $\theta = \phi - y$, we have

$$\dot{\theta} + A\theta + B\theta = B\phi, \quad \theta(0) = 0.$$

Taking the scalar product with θ, since A is skew-adjoint, we get $\langle \dot{\theta} + B\theta, \theta \rangle_X = \langle B\phi, \theta \rangle_X$, and therefore

$$\dot{E}_\theta + \langle B\theta, \theta \rangle_X = \langle B\phi, \theta \rangle_X.$$

Since $\langle B\theta, \theta \rangle_X = \|B^{1/2}\theta\|_X \geq 0$, it follows that $\dot{E}_\theta \leq \langle B\phi, \theta \rangle_X$. As before we apply $\int_0^T \int_0^t$ and hence, since $E_\theta(0) = 0$,

$$\int_0^T E_\theta(t)\,dt \leq \int_0^T \int_0^t \langle B\phi(s), \theta(s) \rangle_X \, ds\, dt = \int_0^T (T-t)\langle B^{1/2}\phi(t), B^{1/2}\theta(t) \rangle_X \, dt.$$

Thanks to the Young inequality, we get, exactly as before,

$$\frac{1}{2}\int_0^T \|\theta(t)\|_X^2\, dt \leq T\|B^{1/2}\| \int_0^T \|B^{1/2}\phi(t)\|_X \|\theta(t)\|_X \, dt$$

$$\leq T^2\|B^{1/2}\|_X^2 \int_0^T \|B^{1/2}\phi(t)\|_X^2 \, dt + \frac{1}{4}\int_0^T \|\theta(t)\|_X^2 \, dt,$$

and finally,

$$\int_0^T \|\theta(t)\|_X^2\, dt \leq 4T^2\|B^{1/2}\|_X^2 \int_0^T \|B^{1/2}\phi(t)\|_X^2 \, dt.$$

Now, since $y = \phi - \theta$, it follows that

$$\int_0^T \|B^{1/2}y(t)\|_X^2\, dt \leq 2\int_0^T \|B^{1/2}\phi(t)\|_X^2\, dt + 2\int_0^T \|B^{1/2}\theta(t)\|_X^2 \, dt$$

$$\leq (2 + 8T^2\|B^{1/2}\|^4) \int_0^T \|B^{1/2}\phi(t)\|_X^2 \, dt.$$

Now, using (2.12) and noting that $E_y(0) = E_\phi(0)$, we infer that

$$C_2 E_\phi(0) \leq (2 + 8T^2\|B^{1/2}\|^4) \int_0^T \|B^{1/2}\phi(t)\|_X^2 \, dt.$$

This is the desired observability inequality. $\qquad\square$

Remark 2.4 This result says that the observability property for the linear conservative equation (2.10) is equivalent to the exponential stability property for the linear damped equation (2.11). This result has been written in [23] for second-order equations, but the proof works exactly in the same way for more general first-order systems, as shown here. More precisely, the statement done in [23] for second-order equations looks as follows:

We have equivalence of:

1. *There exist $T > 0$ and $C > 0$ such that every solution of*

$$\ddot{\phi}(t) + A\phi(t) = 0 \qquad \text{(conservative equation)}$$

satisfies

$$\|A^{1/2}\phi(0)\|_X^2 + \|\dot{\phi}(0)\|_X^2 \le C \int_0^{T_0} \|B^{1/2}\dot{\phi}(t)\|_X^2 \, dt.$$

2. *There exist $C_1 > 0$ and $\delta > 0$ such that every solution of*

$$\ddot{y}(t) + Ay(t) + B\dot{y}(t) = 0 \qquad \text{(damped equation)}$$

satisfies

$$E_y(t) \le C_1 E_y(0)e^{-\delta t},$$

where

$$E_y(t) = \frac{1}{2}\left(\|A^{1/2}y(t)\|_X^2 + \|\dot{y}(t)\|_X^2\right).$$

Remark 2.5 A second remark is that the proof uses in a crucial way the fact that the operator B is bounded. We refer to [5] for a generalization for unbounded operators with degree of unboundedness $\le 1/2$ (i.e., $B \in L(U, D(A^{1/2})')$), and only for second-order equations, with a proof using Laplace transforms, and under a condition on the unboundedness of B that is not easy to check (related to "hidden regularity" results), namely,

$$\forall \beta > 0 \qquad \sup_{\text{Re}(\lambda)=\beta} \|B^*\lambda(\lambda^2 I + A)^{-1}B\|_{L(U)} < +\infty.$$

For instance this works for waves with a nonlocal operator B corresponding to a Dirichlet condition, in the state space $L^2 \times H^{-1}$, but not for the usual Neumann one, in the state space $H^1 \times L^2$ (except in 1D).

2.3.1.2 Semilinear Case

In the case with a nonlinear feedback, still in order to be in agreement with standard notations used in the existing literature, we rather write the equation in the form $\dot{u} + Au + F(u) = 0$, where u now designates the solution (and not the control).

Therefore, from now on and throughout the rest of this chapter, we consider the differential system

$$u'(t) + Au(t) + BF(u(t)) = 0, \qquad (2.13)$$

with $A: D(A) \to X$ a densely defined skew-adjoint operator, $B: X \to X$ a nontrivial bounded self-adjoint nonnegative operator, and $F: X \to X$ a (nonlinear) mapping assumed to be Lipschitz continuous on bounded subsets of X. These are the framework and notations adopted in [4].

If $F = 0$ then the system (2.13) is purely conservative, and $\|u(t)\|_X = \|u(0)\|_X$ for every $t \geq 0$. If $F \neq 0$ then the system (2.13) is expected to be dissipative if the nonlinearity F has "the good sign." Along any solution of (2.13) (while it is well defined), the derivative with respect to time of the energy $E_u(t) = \frac{1}{2}\|u(t)\|_X^2$ is

$$E_u'(t) = -\langle u(t), BF(u(t)) \rangle_X = -\langle B^{1/2} u(t), B^{1/2} F(u(t)) \rangle_X.$$

In the sequel, we will make appropriate assumptions on B and on F ensuring that $E_u'(t) \leq 0$. It is then expected that the solutions are globally well-defined and that their energy decays asymptotically to 0 as $t \to +\infty$.

We make the following assumptions.

- For every $u \in X$

$$\langle u, BF(u) \rangle_X \geq 0.$$

This assumption implies that $E_u'(t) \leq 0$.

The spectral theorem applied to the bounded nonnegative self-adjoint operator B implies that B is unitarily equivalent to a multiplication: there exist a probability space (Ω, μ), a real-valued bounded nonnegative measurable function b defined on X, and an isometry U from $L^2(\Omega, \mu)$ into X, such that $U^{-1} B U f = bf$ for every $f \in L^2(\Omega, \mu)$.

Now, we define the (nonlinear) mapping $\rho: L^2(\Omega, \mu) \to L^2(\Omega, \mu)$ by

$$\rho(f) = U^{-1} F(Uf).$$

We make the following assumptions on ρ:

- $\rho(0) = 0$ and $f\rho(f) \geq 0$ for every $f \in L^2(\Omega, \mu)$.
- There exist $c_1 > 0$ and $c_2 > 0$ such that, for every $f \in L^\infty(\Omega, \mu)$,

$$c_1 g(|f(x)|) \leq |\rho(f)(x)| \leq c_2 g^{-1}(|f(x)|) \quad \text{for almost every } x \in \Omega \text{ such that } |f(x)| \leq 1,$$

$$c_1 |f(x)| \leq |\rho(f)(x)| \leq c_2 |f(x)| \quad \text{for almost every } x \in \Omega \text{ such that } |f(x)| \geq 1,$$

where g is an increasing odd function of class C^1 such that $g(0) = g'(0) = 0$, $sg'(s)^2/g(s) \to 0$ as $s \to 0$, and such that the function H defined by $H(s) = \sqrt{s} g(\sqrt{s})$, for every $s \in [0, 1]$, is strictly convex on $[0, s_0^2]$ for some $s_0 \in (0, 1]$.

This assumption is issued from [2] where the optimal weight method has been developed.

Examples of such functions g are given by

$$g(s) = s/\ln^p(1/s), \quad s^p, \quad e^{-1/s^2}, \quad s^p \ln^q(1/s), \quad e^{-\ln^p(1/s)}.$$

We define the function $\widehat{H}$ on $\mathbb{R}$ by $\widehat{H}(s) = H(s)$ for every $s \in [0, s_0^2]$ and by $\widehat{H}(s) = +\infty$ otherwise. We define the function L on $[0, +\infty)$ by $L(0) = 0$ and, for $r > 0$, by

$$L(r) = \frac{\widehat{H}^*(r)}{r} = \frac{1}{r} \sup_{s \in \mathbb{R}} \left(rs - \widehat{H}(s) \right),$$

where $\widehat{H}^*$ is the convex conjugate of $\widehat{H}$. By construction, the function $L : [0, +\infty) \to [0, s_0^2)$ is continuous and increasing.

We define $\Lambda_H : (0, s_0^2] \to (0, +\infty)$ by $\Lambda_H(s) = H(s)/sH'(s)$, and we set

$$\forall s \geq 1/H'(s_0^2) \qquad \psi(s) = \frac{1}{H'(s_0^2)} + \int_{1/s}^{H'(s_0^2)} \frac{1}{v^2(1 - \Lambda_H((H')^{-1}(v)))} \, dv.$$

The function $\psi : [1/H'(s_0^2), +\infty) \to [0, +\infty)$ is continuous and increasing.

Hereafter, we use the notations $\lesssim$ and $\simeq$ in the estimates, with the following meaning. Let $\mathcal{S}$ be a set, and let F and G be nonnegative functions defined on $\mathbb{R} \times \Omega \times \mathcal{S}$. The notation $F \lesssim G$ (equivalently, $G \gtrsim F$) means that there exists a constant $C > 0$, only depending on the function g or on the mapping ρ, such that $F(t, x, \lambda) \leq CG(t, x, \lambda)$ for all $(t, x, \lambda) \in \mathbb{R} \times \Omega \times \mathcal{S}$. The notation $F_1 \simeq F_2$ means that $F_1 \lesssim F_2$ and $F_1 \gtrsim F_2$.

In the sequel, we choose $\mathcal{S} = X$, or equivalently, using the isometry U, we choose $\mathcal{S} = L^2(\Omega, \mu)$, so that the notation $\lesssim$ designates an estimate in which the constant does not depend on $u \in X$, or on $f \in L^2(\Omega, \mu)$, but depends only on the mapping ρ. We will use these notations to provide estimates on the solutions $u(\cdot)$ of (2.13), meaning that the constants in the estimates do not depend on the solutions.

Theorem 2.6 [4] *In addition to the above assumptions, we assume that there exist $T > 0$ and $C_T > 0$ such that*

$$C_T \|\phi(0)\|_X^2 \leq \int_0^T \|B^{1/2}\phi(t)\|_X^2 \, dt,$$

for every solution of $\phi'(t) + A\phi(t) = 0$ (observability inequality for the linear conservative equation).

Table 2.1. *Examples*

$g(s)$	$\Lambda_H(s)$	decay of $E(t)$
$s/\ln^p(1/s), p > 0$	$\limsup\limits_{x \searrow 0} \Lambda_H(s) = 1$	$e^{-t^{1/(p+1)}}/t^{1/(p+1)}$
s^p on $[0, s_0^2], p > 1$	$\Lambda_H(s) \equiv \frac{2}{p+1} < 1$	$t^{-2/(p-1)}$
e^{-1/s^2}	$\lim\limits_{s \searrow 0} \Lambda_H(s) = 0$	$1/\ln(t)$
$s^p \ln^q(1/s), p > 1, q > 0$	$\lim\limits_{s \searrow 0} \Lambda_H(s) = \frac{2}{p+1} < 1$	$t^{-2/(p-1)} \ln^{-2q/(p-1)}(t)$
$e^{-\ln^p(1/s)}, p > 1$	$\lim\limits_{s \searrow 0} \Lambda_H(s) = 0$	$e^{-2\ln^{1/p}(t)}$

Then, for every $u_0 \in X$, there exists a unique solution $u(\cdot) \in C^0(0, +\infty; X) \cap C^1(0, +\infty; D(A)')$ of (2.13) such that $u(0) = u_0$.[2] Moreover, the energy of any solution satisfies

$$E_u(t) \lesssim T \max(\gamma_1, E_u(0)) L\left(\frac{1}{\psi^{-1}(\gamma_2 t)}\right),$$

for every time $t \geq 0$, with $\gamma_1 \simeq \|B\|/\gamma_2$ and $\gamma_2 \simeq C_T/T(T^2\|B^{1/2}\|^4 + 1)$. If moreover

$$\limsup_{s \searrow 0} \Lambda_H(s) < 1, \tag{2.14}$$

then we have the simplified decay rate

$$E_u(t) \lesssim T \max(\gamma_1, E_u(0)) (H')^{-1}\left(\frac{\gamma_3}{t}\right),$$

for every time $t \geq 0$, for some positive constant $\gamma_3 \simeq 1$.

Note the important fact that this result gives *sharp decay rates* (see Table 2.1 for examples).

Theorem 2.6 improves and generalizes to a wide class of equations the main result of [3], in which the authors dealt with locally damped wave equations. Examples of applications are given in [4], that we mention here without giving the precise framework, assumptions, and comments:

- Schrödinger equation with nonlinear damping (nonlinear absorption):

$$i\partial_t u(t, x) + \triangle u(t, x) + ib(x)u(t, x)\rho(x, |u(t, x)|) = 0.$$

---

[2] Here, the solution is understood in the weak sense (see [10, 18]), and $D(A)'$ is the dual of $D(A)$ with respect to the pivot space X. If $u_0 \in D(A)$, then $u(\cdot) \in C^0(0, +\infty; D(A)) \cap C^1(0, +\infty; X)$.

- Wave equation with nonlinear damping:

$$\partial_{tt}u(t,x) - \Delta u(t,x) + b(x)\rho(x, \partial_t u(t,x)) = 0.$$

- Plate equation with nonlinear damping:

$$\partial_{tt}u(t,x) + \Delta^2 u(t,x) + b(x)\rho(x, \partial_t u(t,x)) = 0.$$

- Transport equations with nonlinear damping:

$$\partial_t u(t,x) + \mathrm{div}(v(x)u(t,x)) + b(x)\rho(x, u(t,x)) = 0, \qquad x \in \mathbb{T}^n,$$

with $\mathrm{div}(v) = 0$.
- Dissipative equations with nonlocal terms:

$$\partial_t f + v \cdot \nabla_x f = \rho(f),$$

with kernels ρ satisfying the sign assumption $f\rho(f) \geq 0$.

Proof It is interesting to quickly give the main steps of the proof.

- **Step 1**. Comparison of the nonlinear equation with the linear damped model:

  Prove that the solutions of

$$u'(t) + Au(t) + BF(u(t)) = 0,$$

$$z'(t) + Az(t) + Bz(t) = 0, \qquad z(0) = u(0),$$

  satisfy

$$\int_0^T \|B^{1/2}z(t)\|_X^2\, dt \leq 2 \int_0^T \left(\|B^{1/2}u(t)\|_X^2 + \|B^{1/2}F(u(t))\|_X^2 \right) dt.$$

- **Step 2**. Comparison of the linear damped equation with the conservative linear equation:

  Prove that the solutions of

$$z'(t) + Az(t) + Bz(t) = 0, \qquad z(0) = u(0),$$

$$\phi'(t) + A\phi(t) = 0, \qquad \phi(0) = u(0),$$

  satisfy

$$\int_0^T \|B^{1/2}\phi(t)\|_X^2\, dt \leq k_T \int_0^T \|B^{1/2}z(t)\|_X^2\, dt,$$

with $k_T = 8T^2 \|B^{1/2}\|^4 + 2$.

- **Step 3**. Following the optimal weight method introduced by F. Alabau (see, e.g., [2]), we set $w(s) = L^{-1}\left(\frac{s}{\beta}\right)$ with β appropriately chosen. Nonlinear energy estimate:

  Prove that

  $$\int_0^T w(E_\phi(0))\left(\|B^{1/2}u(t)\|_X^2 + \|B^{1/2}F(u(t))\|_X^2\right) dt$$

  $$\lesssim T\|B\|H^*(w(E_\phi(0))) + (w(E_\phi(0)) + 1)\int_0^T \langle Bu(t), F(u(t))\rangle_X \, dt.$$

- **Step 4**. End of the proof:

  Using the results of the three steps above, we have

  $$T\|B\|H^*(w(E_\phi(0))) + (w(E_\phi(0)) + 1)\int_0^T \langle Bu(t), F(u(t))\rangle_X \, dt$$

  $$\gtrsim \int_0^T w(E_\phi(0))\left(\|B^{1/2}u(t)\|_X^2 + \|B^{1/2}F(u(t))\|_X^2\right) dt \qquad \text{(Step 3)}$$

  $$\gtrsim \int_0^T w(E_\phi(0))\|B^{1/2}z(t)\|_X^2 \, dt \qquad \text{(Step 1)}$$

  $$\gtrsim \int_0^T w(E_\phi(0))\|B^{1/2}\phi(t)\|_X^2 \, dt \qquad \text{(Step 2)}$$

  $$\gtrsim \text{Cst } w(E_\phi(0))E_\phi(0) \qquad \text{(uniform observability inequality)}$$

  from which we infer that

  $$E_u(T) \le E_u(0)\left(1 - \rho_T L^{-1}\left(\frac{E_u(0)}{\beta}\right)\right),$$

  and then the exponential decrease is finally established. $\qquad \square$

2.3.2 Space Semidiscretizations

In this section, we define a general space semidiscrete version of (2.13), with the objective of obtaining a theorem similar to Theorem 2.6, but in this semidiscrete setting, with estimates that are uniform with respect to the mesh parameter. As we are going to see, uniformity is not true in general, and in order to recover it we add an appropriate extra numerical viscosity term in the numerical scheme.

2.3.2.1 Space Semidiscretization Setting

We denote by $\triangle x > 0$ the space discretization parameter (typically, step size of the mesh), with $0 < \triangle x < \triangle x_0$, for some fixed $\triangle x_0 > 0$. We follow [28, 29] for the setting. Let $(X_{\triangle x})_{0 < \triangle x < \triangle x_0}$ be a family of finite-dimensional vector spaces ($X_{\triangle x} \sim \mathbb{R}^{N(\triangle x)}$ with $N(\triangle x) \in \mathbb{N}$).

We use the notations $\lesssim$ and $\simeq$ as before, also meaning that the involved constants are uniform with respect to $\triangle x$.

Let $\beta \in \rho(A)$ (resolvent of A). Following [18], we define $X_{1/2} = (\beta \mathrm{id}_X - A)^{-1/2}(X)$, endowed with the norm $\|u\|_{X_{1/2}} = \|(\beta \mathrm{id}_X - A)^{1/2} u\|_X$ (for instance, if $A^{1/2}$ is well defined, then $X_{1/2} = D(A^{1/2})$), and we define $X_{-1/2} = X'_{1/2}$ (dual with respect to X).

The general semidiscretization setting is the following. We assume that, for every $\triangle x \in (0, \triangle x_0)$, there exist linear mappings $P_{\triangle x} : X_{-1/2} \to X_{\triangle x}$ and $\widetilde{P}_{\triangle x} : X_{\triangle x} \to X_{1/2}$ such that $P_{\triangle x} \widetilde{P}_{\triangle x} = \mathrm{id}_{X_{\triangle x}}$, and such that $P_{\triangle x} = \widetilde{P}^*_{\triangle x}$. We assume that the scheme is convergent, that is, $\|(I - \widetilde{P}_{\triangle x} P_{\triangle x}) u\|_X \to 0$ as $\triangle x \to 0$, for every $u \in X$. Here, we have implicitly used the canonical injections $D(A) \hookrightarrow X_{1/2} \hookrightarrow X \hookrightarrow X_{-1/2}$ (see [18]).

For every $\triangle x \in (0, \triangle x_0)$:

- $X_{\triangle x}$ is endowed with the Euclidean norm $\|u_{\triangle x}\|_{\triangle x} = \|\widetilde{P}_{\triangle x} u_{\triangle x}\|_X$, for $u_{\triangle x} \in X_{\triangle x}$. The corresponding scalar product is denoted by $\langle \cdot, \cdot \rangle_{\triangle x}$.[3]
- We set[4]

$$A_{\triangle x} = P_{\triangle x} A \widetilde{P}_{\triangle x}, \qquad B_{\triangle x} = P_{\triangle x} B \widetilde{P}_{\triangle x}.$$

Since $P_{\triangle x} = \widetilde{P}^*_{\triangle x}$, $A_{\triangle x}$ is skew-symmetric and $B_{\triangle x}$ is symmetric nonnegative

- We define $F_{\triangle x} : X_{\triangle x} \to X_{\triangle x}$ by

$$\forall u_{\triangle x} \in X_{\triangle x} \qquad F_{\triangle x}(u_{\triangle x}) = P_{\triangle x} F(\widetilde{P}_{\triangle x} u_{\triangle x}).$$

Note that $B_{\triangle x}$ is uniformly bounded with respect to $\triangle x$, and $F_{\triangle x}$ is Lipschitz continuous on bounded subsets of $X_{\triangle x}$, uniformly with respect to $\triangle x$.

Now, a priori we consider the space semidiscrete approximation of (2.13) given by

$$u'_{\triangle x}(t) + A_{\triangle x} u_{\triangle x}(t) + B_{\triangle x} F_{\triangle x}(u_{\triangle x}(t)) = 0, \qquad (2.15)$$

and we wonder whether we are able or not to ensure a uniform decay rate of solutions of (2.15).

It happens that the answer is **NO** in general. Let us give hereafter an example in the linear case ($F = \mathrm{id}_X$).

---

[3] Note that $\|\widetilde{P}_{\triangle x}\|_{L(X_{\triangle x}, X)} = 1$ and that, by the Uniform Boundedness Principle, $\|P_{\triangle x}\|_{L(X, X_{\triangle x})} \lesssim 1$.

[4] We implicitly use the canonical extension of the operator $A : X_{1/2} \to X_{-1/2}$.

2.3.2.2 Finite-Difference Space Semidiscretization
of the 1D Damped Wave Equation

Let us consider the 1D damped Dirichlet wave equation

$$y_{tt} - y_{xx} + ay_t = 0, \quad 0 < x < 1, \ t > 0,$$
$$y(t,0) = y(t,1) = 0, \tag{2.16}$$

with $a \geq 0$ measurable and bounded, satisfying $a(x) \geq \alpha > 0$ on $\omega \subset (0,1)$ of measure $0 < |\omega| < 1$. This equation fits in our framework with

$$u = \begin{pmatrix} y \\ y_t \end{pmatrix}, \quad A = \begin{pmatrix} 0 & \mathrm{id} \\ -\triangle & 0 \end{pmatrix}, \quad B = a\,\mathrm{id}, \quad F = \mathrm{id}.$$

The energy along any solution is

$$E(t) = \frac{1}{2} \int_0^1 \left(y_t(t,x)^2 + y_x(t,x)^2 \right) dx,$$

and we have

$$E'(t) = - \int_0^1 a(x) y_t(t,x)^2 \, dx \leq 0.$$

It is known that $E(t)$ has an exponential decrease. Indeed, by Theorem 2.3, the exponential decrease is equivalent to the observability inequality

$$\int_0^T \int_\omega \phi_t(t,x)^2 \, dx \, dt \geq C \int_0^1 (\phi_t(0,x)^2 + \phi(0,x)^2) \, dx \tag{2.17}$$

for every solution ϕ of the conservative equation

$$\phi_{tt} - \phi_{xx} = 0, \quad \phi(t,0) = \phi(t,1) = 0, \tag{2.18}$$

and we have the following result.

Proposition 2.7 The observability inequality (2.17) holds, for every $T \geq 2$.

Proof The proof is elementary and can be done by a simple spectral expansion. Any solution ϕ of (2.18) can be expanded as

$$\phi(t,x) = \sum_{k=1}^\infty \left(a_k \cos(k\pi t) + \frac{b_k}{k\pi} \sin(k\pi t) \right) \sin(k\pi x),$$

with

$$\phi(0,x) = \sum_{k=1}^\infty a_k \sin(k\pi x), \quad \text{and} \quad \phi_t(0,x) = \sum_{k=1}^\infty b_k \sin(k\pi x).$$

Then,

$$\int_0^1 (\phi_t(0,x)^2 + \phi(0,x)^2)\, dx = \frac{1}{2}\sum_{k=1}^{\infty}(a_k^2 + b_k^2).$$

Now, for every $T \geq 2$, we have

$$\int_0^T \int_\omega \phi_t(t,x)^2\, dx\, dt \geq \int_0^2 \int_\omega \phi_t(t,x)^2\, dx\, dt,$$

and

$$\int_0^2 \int_\omega \phi_t(t,x)^2\, dx\, dt = \sum_{j,k=1}^{\infty}\int_0^2 (a_j\cos(j\pi t) + b_j\sin(j\pi t))(a_k\cos(k\pi t)$$

$$+\, b_k\sin(k\pi t))\, dt \times \int_\omega \sin(j\pi x)\sin(k\pi x)\, dx$$

$$= \sum_{j=1}^{\infty}(a_j^2 + b_j^2)\int_\omega \sin^2(j\pi x)\, dx.$$

At this step, we could simply conclude by using the fact that the sequence of functions $\sin^2(j\pi x)$ converges weakly to $1/2$, and then infer an estimate from below. But it is interesting to note that we can be more precise by showing the following simple result, valid for any measurable subset ω of $[0,1]$.

Lemma 2.8 For every measurable subset $\omega \subset [0,1]$, there holds

$$\forall j \in \mathbb{N}^* \qquad \int_\omega \sin^2(j\pi x)\, dx \geq \frac{|\omega| - \sin|\omega|}{2},$$

where $|\omega|$ is the Lebesgue measure of ω.

With this lemma, the observability property follows and we are done. It remains to prove the lemma.

For a *fixed* integer j, let us consider the problem of minimizing the functional

$$K_j(\omega') = \int_{\omega'} \sin^2(j\pi x)\, dx$$

over the set of all measurable subsets $\omega' \subset [0,1]$ of Lebesgue measure $|\omega'| = |\omega|$. Clearly there exists a unique (up to zero measure subsets) optimal set, characterized as a level set of the function $x \mapsto \sin^2(j\pi x)$, which is the set

$$\omega_j^{\inf} = \left(0, \frac{|\omega|}{2j}\right) \cup \bigcup_{k=1}^{j-1}\left(\frac{k}{j} - \frac{|\omega|}{2j}, \frac{k}{j} + \frac{|\omega|}{2j}\right) \cup \left(1 - \frac{|\omega|}{2j}, \pi\right),$$

and there holds

$$\int_{\omega_j^{\inf}} \sin^2(j\pi x)\,dx = 2j \int_0^{|\omega|/2j} \sin^2 jx\,dx = 2 \int_0^{|\omega|/2} \sin^2 u\,du = \frac{1}{2}(|\omega| - \sin(|\omega|)).$$

Since the quantity is independent of j, the lemma follows, and thus the proposition as well. □

Hence, at this step, we know that the energy is exponentially decreasing for the continuous damped wave equation.

Let us now consider the usual finite-difference space semidiscretization model:

$$\ddot{y}_j - \frac{y_{j+1} - 2y_j + y_{j-1}}{(\triangle x)^2} + a_j \dot{y}_j = 0, \qquad j = 1, \ldots, N,$$

$$y_0(t) = y_{N+1}(t) = 0.$$

The energy of any solution is given by

$$E_{\triangle x}(t) = \frac{\triangle x}{2} \sum_{j=0}^{N} \left((y_j'(t))^2 + \left(\frac{y_{j+1}(t) - y_j(t)}{\triangle x} \right)^2 \right),$$

and we have

$$E_{\triangle x}'(t) = -\triangle x \sum_{j=1}^{N} a_j (y_j'(t))^2 \leq 0.$$

We claim that $E_{\triangle x}(t)$ does not decrease uniformly exponentially.
Indeed, otherwise there would hold

$$E_{\triangle x}(0) \leq C \triangle x \sum_{j=1}^{N} \int_0^T a_j(\dot{\phi}_j(t))^2\,dt$$

for every solution of the conservative system

$$\ddot{\phi}_j - \frac{\phi_{j+1} - 2\phi_j + \phi_{j-1}}{(\triangle x)^2} = 0, \qquad j = 1, \ldots, N,$$

$$\phi_0(t) = \phi_{N+1}(t) = 0, \tag{2.19}$$

for some time $T > 0$ and some *uniform* $C > 0$. But this is known to be wrong. To be more precise, it is true that, for every fixed value of $\triangle x$, there exists a constant $C_{\triangle x}$, depending on $\triangle x$, such that any solution of (2.19) satisfies

$$E_{\triangle x}(0) \leq C_{\triangle x} \triangle x \sum_{j=1}^{N} \int_0^T a_j(\dot{\phi}_j(t))^2\,dt.$$

This is indeed the usual observability inequality in finite dimension, but with a constant $C_{\triangle x}$ depending on $\triangle x$ (by the way, this observability inequality holds true for any $T > 0$, but we are of course interested in taking $T \geq 2$ in the hope of being able to pass to the limit). Now, the precise negative result is the following.

Proposition 2.9 [33] We have

$$C_{\triangle x} \xrightarrow[\triangle x \to 0]{} +\infty.$$

In other words, the observability constant is not uniform and observability is lost when we pass to the limit. The proof of this proposition combines the following facts:

- Using *gaussian beams* it can be shown that along every bicharacteristic ray there exists a solution of the wave equation whose energy is *localized along this ray*.
- The *velocity of high-frequency wave packets* for the discrete model *tends to 0* as $\triangle x$ tends to 0.

Then, for every $T > 0$, for $\triangle x > 0$ small enough there exist initial data whose corresponding solution is concentrated along a ray that does not reach the observed region ω within time T. The result follows.

Remark 2.10 This proof of nonuniformity is not easy in the case of an internal observation, and relies on microlocal considerations. It can be noticed that nonuniformity is much easier to prove, in an elementary way, in the case of a boundary observation. We reproduce here the main steps of [24] with easy-to-do computations:

1. The eigenelements of the (finite-difference) matrix $\triangle_{\triangle x}$ are

$$e_{\triangle x, j} = \begin{pmatrix} \vdots \\ \sin(j\pi k \triangle x) \\ \vdots \end{pmatrix}, \quad \lambda_{\triangle x, j} = -\frac{4}{(\triangle x)^2} \sin^2 \frac{j\pi \triangle x}{2}, \quad j = 1, \dots, N.$$

2. If $\triangle x \to 0$ then

$$\sqrt{|\lambda_{\triangle x, n}|} - \sqrt{|\lambda_{\triangle x, n-1}|} \sim \pi^2 \triangle x.$$

3. The vector-valued function of time

$$\phi_{\triangle x}(t) = \frac{1}{\sqrt{|\lambda_{\triangle x, N}|}} \left(e^{it\sqrt{|\lambda_{\triangle x, N}|}} e_{\triangle x, N} - e^{it\sqrt{|\lambda_{\triangle x, N-1}|}} e_{\triangle x, N-1} \right)$$

is solution of $\ddot{\phi}_{\triangle x}(t) = \triangle_{\triangle x} \phi_{\triangle x}(t)$, and if $\triangle x$ tends to 0 then

- the term $\triangle x \sum_{j=0}^{N} \left(\left| \dfrac{\phi_{j+1}(0) - \phi_j(0)}{\triangle x} \right|^2 + |\dot{\phi}_j(0)|^2 \right)$ has the order of 1,

- the term $\triangle x \displaystyle\int_0^T \left(\dfrac{\dot{\phi}_N(t)}{\triangle x} \right)^2 dt$ has the order of $\triangle x$.

4. Therefore $C_{\triangle x} \to +\infty$ as $\triangle x \to 0$.

Remedy for this example. A remedy to this lack of uniformity has been proposed in [40], which consists of adding in the numerical scheme a suitable extra numerical viscosity term, as follows:

$$\ddot{y}_j - \frac{y_{j+1} - 2y_j + y_{j-1}}{(\triangle x)^2} + a_j \dot{y}_j - (\triangle x)^2 \left(\frac{\dot{y}_{j+1} - 2\dot{y}_j + \dot{y}_{j-1}}{(\triangle x)^2} \right) = 0, \quad j = 1, \ldots, N$$

$$y_0(t) = y_{N+1}(t) = 0.$$

With matrix notations, this can be written as

$$\ddot{y}_{\triangle x} - \triangle_{\triangle x} y_{\triangle x} + a_{\triangle x} \dot{y}_{\triangle x} - (\triangle x)^2 \triangle_{\triangle x} \dot{y}_{\triangle x} = 0.$$

Now the energy decreases according to

$$\dot{E}_{\triangle x} = -\triangle x \sum_{j=1}^{N} a_j \dot{y}_j^2 - (\triangle x)^3 \sum_{j=0}^{N} \left(\frac{\dot{y}_{j+1} - \dot{y}_j}{\triangle x} \right)^2,$$

and we have the following result.

Proposition 2.11 [40] There exist $C > 0$ and $\delta > 0$ such that, for every $\triangle x$,

$$E_{\triangle x}(t) \le C \exp(-\delta t) E_{\triangle x}(0).$$

In other words, adding an appropriate viscosity term allows to recover a uniform exponential decrease.

An intuitive explanation is that the high frequency eigenvalues are of the order of $\frac{1}{(\triangle x)^2}$. Indeed,

$$\lambda_{\triangle x, j}^2 = -\frac{4}{(\triangle x)^2} \sin^2 \frac{j\pi \triangle x}{2}, \qquad j = 1, \ldots, N.$$

Then for high frequencies the viscosity term $(\triangle x)^2 \triangle_{\triangle x} \dot{y}_{\triangle x}$ has the order of $\dot{y}_{\triangle x}$, and hence it behaves as a *damping on high frequencies*. The main result of [40] shows that it suffices to recover a uniform exponential decay.

Their proof mainly consists of proving the *uniform observability inequality*

$$E_{\triangle x}(0) \le C \triangle x \sum_{j=1}^{N} \int_0^T \left(a_j (\dot{\phi}_j(t))^2 + (\triangle x)^2 \left(\frac{\dot{\phi}_{j+1} - \dot{\phi}_j}{\triangle x} \right)^2 \right) dt,$$

for every solution of

$$\frac{\ddot{\phi}_{j+1} + 2\ddot{\phi}_j + \ddot{\phi}_{j-1}}{4} - \frac{\phi_{j+1} - 2\phi_j + \phi_{j-1}}{(\triangle x)^2} = 0, \qquad j = 1, \ldots, N,$$

$$\phi_0(t) = \phi_{N+1}(t) = 0,$$

and this follows from a meticulous application of a discrete version of the multiplier method.

Note that they also prove the result in a 2D square, and that this was generalized to any regular 2D domain in [35].

2.3.2.3 General Results for Linear Damped Equations, with a Viscosity Operator

The idea of adding a viscosity in the numerical scheme, used in [40] in order to recover a uniform exponential decay rate for the finite-difference space semidiscretization of the linear damped wave equation, can be generalized as follows.

We consider (2.13) in the linear case $F = \text{id}_X$, that is,

$$u'(t) + Au(t) + Bu(t) = 0,$$

with $A : D(A) \to X$ skew-adjoint and $B = B^* \geq 0$ bounded. The energy along any solution is

$$E(t) = \frac{1}{2} \|u(t)\|_X^2,$$

and we have

$$\dot{E}(t) = -\|B^{1/2}u(t)\|_X^2.$$

We consider the space semidiscrete model *with viscosity*:

$$\dot{u}_{\triangle x} + A_{\triangle x}u_{\triangle x} + B_{\triangle x}u_{\triangle x} + (\triangle x)^\sigma V_{\triangle x}u_{\triangle x} = 0, \qquad (2.20)$$

where $V_{\triangle x} : X_{\triangle x} \to X_{\triangle x}$ is a positive self-adjoint operator, called the *viscosity operator*.

In the previous example (1D damped wave equation), the viscosity term was $-(\triangle x)^2 \triangle_{\triangle x}y'_{\triangle x}$, hence

$$\sigma = 2, \qquad V_{\triangle x} = \begin{pmatrix} 0 & 0 \\ 0 & -\triangle_{\triangle x} \end{pmatrix}.$$

Note that it is not positive definite, however it was proved to be sufficient to infer uniformity.

In general, as we are going to see, a typical example of a viscosity operator is $V_{\triangle x} = \sqrt{A^*_{\triangle x}A_{\triangle x}}$, plus, possibly, $\varepsilon\,\text{id}_{X_{\triangle x}}$ (to make it positive).

In any case, the energy along a solution of (2.20), defined by

$$E_{u_{\triangle x}}(t) = \frac{1}{2}\|u_{\triangle x}(t)\|^2_{\triangle x},$$

has a derivative in time given by

$$\dot{E}_{u_{\triangle x}}(t) = -\|B^{1/2}_{\triangle x}u_{\triangle x}(t)\|^2_{\triangle x} - (\triangle x)^\sigma \|(\mathcal{V}_{\triangle x})^{1/2}u_{\triangle x}(t)\|^2_{\triangle x} \leq 0.$$

Now, we have the following result.

Theorem 2.12 [21] *Assume that the uniform observability inequality*

$$C\|\phi_{\triangle x}(0)\|^2 \leq \int_0^T \left(\|B^{1/2}_{\triangle x}\phi_{\triangle x}(t)\|^2 + (\triangle x)^\sigma\|\mathcal{V}^{1/2}_{\triangle x}\phi_{\triangle x}(t)\|^2\right) dt$$

holds for every solution of the conservative system

$$\dot{\phi}_{\triangle x} + A_{\triangle x}\phi_{\triangle x} = 0.$$

Then the uniform exponential decrease holds, i.e., there exist $C > 0$ and $\delta > 0$ such that

$$E_{u_{\triangle x}}(t) \leq C\exp(-\delta t)E_{u_{\triangle x}}(0),$$

for every $\triangle x$, for every solution $u_{\triangle x}$ of (2.20).

This theorem says that, in general, adding a viscosity term in the numerical scheme helps to recover a uniform exponential decay, provided a uniform observability inequality holds true for the corresponding conservative equation. Of course, given a specific equation, the difficulty is to establish the uniform observability inequality. This is a difficult issue, investigated in some particular cases (see [48] for a survey).

Let us mention several other related results. In [19], such uniform properties are established for second-order equations, with a self-adjoint positive operator, semidiscretized in space by means of finite elements on general meshes, not necessarily regular. The results have been generalized in [34]. In [1], one can find similar results under appropriate spectral gap conditions (in this reference, we can also find results on polynomial decay, see further). In [36, 37], viscosity operators are used for the plate equation, and for general classes of second-order evolution equations under appropriate spectral gap assumptions, combined with the uniform Huang–Prüss conditions derived in [32].

Remark 2.13 In order to recover uniformity, there exist other ways than using viscosities (see [48]):

- Use an appropriate numerical scheme (such as mixed finite elements, multigrids).

  For instance, considering again the 1D damped wave equation (2.16), the mixed finite element space semidiscretization model is

  $$\frac{\ddot{y}_{j+1} + 2\ddot{y}_j + \ddot{y}_{j-1}}{4} - \frac{y_{j+1} - 2y_j + y_{j-1}}{(\triangle x)^2} + \frac{a_{j-1/2}}{4}(\dot{y}_{j-1} + \dot{y}_j)$$

  $$+ \frac{a_{j+1/2}}{4}(\dot{y}_j + \dot{y}_{j+1}) = 0,$$

  $$y_0(t) = y_{N+1}(t) = 0, \qquad j = 1, \ldots, N.$$

  The energy of any solution is

  $$E_{\triangle x}(t) = \frac{\triangle x}{2} \sum_{j=0}^{N} \left(\left(\frac{\dot{y}_{j+1}(t) + \dot{y}_j(t)}{2} \right)^2 + \left(\frac{y_{j+1}(t) - y_j(t)}{\triangle x} \right)^2 \right),$$

  and one has

  $$\dot{E}_{\triangle x}(t) = -\triangle x \sum_{j=0}^{N-1} a_{j+1/2} \left(\frac{\dot{y}_{j+1}(t) + \dot{y}_j(t)}{2} \right)^2 \leq 0.$$

  It is then proved in [20] that $E_{\triangle x}(t)$ decreases exponentially, uniformly with respect to $\triangle x$.

  The proof consists of proving that the *uniform* observability inequality:

  $$E_{\triangle x}(0) \leq C\triangle x \sum_{j=1}^{N} \int_0^T a_{j+1/2} \left(\frac{\dot{\phi}_{j+1}(t) + \dot{\phi}_j(t)}{2} \right)^2 dt,$$

  for any solution of

  $$\ddot{\phi}_j - \frac{\phi_{j+1} - 2\phi_j + \phi_{j-1}}{(\triangle x)^2} = 0, \qquad j = 1, \ldots, N,$$

  $$\phi_0(t) = \phi_{N+1}(t) = 0.$$

  The proof of this uniform observability inequality is done thanks to a careful spectral analysis, using Ingham's inequality.

  The main difference with the finite-difference numerical scheme is that, here, the discrete eigenvalues are given by

  $$\lambda_{\triangle x, j}^2 = -\frac{4}{(\triangle x)^2} \tan^2 \frac{j\pi h}{2}, \qquad j = 1, \ldots, N,$$

and then their behavior at high frequencies (j close to N) is very different from the case of finite differences where the tan function was replaced with the sin function: the derivative of the discrete symbol is now bounded below and hence there are not anymore any wavepackets with vanishing speed (see [48]).

• Filtering high frequencies.

In the general framework

$$\dot{u} + Au + Bu = 0,$$

with $A : D(A) \to X$ skew-adjoint and $B = B^* \geq 0$ bounded, considering the semidiscrete model

$$\dot{u}_{\triangle x} + A_{\triangle x} u_{\triangle x} + B_{\triangle x} u_{\triangle x} = 0,$$

the idea (well surveyed in [48]) consists of filtering out the high frequencies, so that the uniform observability inequality

$$C\|\phi_{\triangle x}(0)\|^2 \leq \int_0^T \|B_{\triangle x}^{1/2} \phi_{\triangle x}(t)\|^2 \, dt$$

holds for every solution of the conservative system $\dot{\phi}_{\triangle x} = A_{\triangle x} \phi_{\triangle x}$ with

$$\phi_{\triangle x}(0) \in \mathcal{C}_{\delta/(\triangle x)^\sigma}(A_{\triangle x}) \qquad \text{(spectrally filtered initial data),}$$

with

$$\mathcal{C}_s(A_{\triangle x}) = \text{Span}\{e_{\triangle x, j} \mid |\lambda_{\triangle x, j}| \leq s\}.$$

Here, the vectors $e_{\triangle x, j}$ are eigenvectors of $A_{\triangle x}$ associated with the eigenvalues $\lambda_{\triangle x, j}$. Then the uniform exponential decrease holds.

Actually, as noticed in [21], we have equivalence of:

– Every solution of $\dot{\phi}_{\triangle x} = A_{\triangle x} \phi_{\triangle x}$ satisfies

$$C_1 \|\phi_{\triangle x}(0)\|^2 \leq \int_0^T \left(\|B_{\triangle x}^{1/2} \phi_{\triangle x}(t)\|^2 + (\triangle x)^\sigma \|V_{\triangle x}^{1/2} \phi_{\triangle x}(t)\|^2 \right) dt.$$

– Every solution of $\dot{\phi}_{\triangle x} = A_{\triangle x} \phi_{\triangle x}$ with $\phi_{\triangle x}(0) \in \mathcal{C}_{\delta/(\triangle x)^\sigma}(A_{\triangle x})$ satisfies

$$C_2 \|\phi_{\triangle x}(0)\|^2 \leq \int_0^T \|B_{\triangle x}^{1/2} \phi_{\triangle x}(t)\|^2 \, dt.$$

In conclusion, there are several ways to recover uniform properties for space semidiscretizations:

1. Add a viscosity term (sometimes called Tychonoff regularization): its role being to damp out the spurious high frequencies.

2. Use an adapted method (like mixed finite elements, or multigrid methods), directly leading to a uniform observability inequality.
3. Filter out high frequencies, by using spectral projections.

In our case, we are going to focus on the use of appropriate viscosity terms, in order to deal more easily with nonlinear terms.

2.3.2.4 Viscosity in General Semilinear Equations

We now come back to the study of the general semilinear equation (2.13), and according to the previous analysis, we consider the space semidiscrete approximation of (2.13) given by

$$u'_{\triangle x}(t) + A_{\triangle x}u_{\triangle x}(t) + B_{\triangle x}F_{\triangle x}(u_{\triangle x}(t)) + (\triangle x)^{\sigma}\mathcal{V}_{\triangle x}u_{\triangle x}(t) = 0. \quad (2.21)$$

As before, the additional term $(\triangle x)^{\sigma}\mathcal{V}_{\triangle x}u_{\triangle x}(t)$, with $\sigma > 0$, is a *numerical viscosity term* whose role is crucial in order to establish decay estimates that are uniform with respect to $\triangle x$. We only assume that $\mathcal{V}_{\triangle x} : X_{\triangle x} \to X_{\triangle x}$ (viscosity operator) is a positive self-adjoint operator.

Our objective is to be able to guarantee a uniform decay, with the decay rate obtained in Theorem 2.6.

The energy of a solution $u_{\triangle x}$ of (2.21) is

$$E_{u_{\triangle x}}(t) = \frac{1}{2}\|u_{\triangle x}(t)\|^2_{\triangle x},$$

and we have, as long as the solution is well-defined,

$$E'_{u_{\triangle x}}(t) = -\langle u_{\triangle x}(t), B_{\triangle x}F_{\triangle x}(u_{\triangle x}(t))\rangle_{\triangle x} - (\triangle x)^{\sigma}\|(\mathcal{V}_{\triangle x})^{1/2}u_{\triangle x}(t)\|^2_{\triangle x}.$$

We assume that

$$\langle u_{\triangle x}, B_{\triangle x}F_{\triangle x}(u_{\triangle x})\rangle_{\triangle x} + (\triangle x)^{\sigma}\|(\mathcal{V}_{\triangle x})^{1/2}u_{\triangle x}\|^2_{\triangle x} \geq 0,$$

for every $u_{\triangle x} \in X_{\triangle x}$. Hence the energy $E_{u_{\triangle x}}(t)$ is nonincreasing.

For every $f \in L^2(\Omega, \mu)$, we set

$$\tilde{\rho}_{\triangle x}(f) = U^{-1}\tilde{P}_{\triangle x}P_{\triangle x}U\rho(f) = U^{-1}\tilde{P}_{\triangle x}P_{\triangle x}F(Uf).$$

The mapping $\tilde{\rho}_{\triangle x}$ is the mapping ρ filtered by the "sampling operator"

$$U^{-1}\tilde{P}_{\triangle x}P_{\triangle x}U = (P_{\triangle x}U)^*P_{\triangle x}U.$$

By assumption, the latter operator converges pointwise to the identity as $\triangle x \to 0$, and in many numerical schemes it corresponds to take sampled values of a given function f.

We have $\tilde{\rho}_{\triangle x}(0) = 0$. Setting $f_{\triangle x} = U^{-1}\tilde{P}_{\triangle x}u_{\triangle x}$, we assume that

$$f_{\triangle x}\tilde{\rho}_{\triangle x}(f_{\triangle x}) \geq 0,$$

and that

$$c_1 g(|f_{\triangle x}(x)|) \le |\tilde{p}_{\triangle x}(f_{\triangle x})(x)| \le c_2 g^{-1}(|f_{\triangle x}(x)|) \quad \text{for a.e. } x \in \Omega \text{ such that } |f_{\triangle x}(x)| \le 1,$$

$$c_1 |f_{\triangle x}(x)| \le |\tilde{p}_{\triangle x}(f_{\triangle x})(x)| \le c_2 |f_{\triangle x}(x)| \quad \text{for a.e. } x \in \Omega \text{ such that } |f_{\triangle x}(x)| \ge 1,$$

for every $u_{\triangle x} \in X_{\triangle x}$, for every $\triangle x \in (0, \triangle x_0)$.

Note that these additional assumptions are valid for many classical numerical schemes, such as finite differences, finite elements, and in more general, for any method based on Lagrange interpolation, in which inequalities or sign conditions are preserved under sampling. But for instance this assumption may fail for spectral methods (global polynomial approximation) in which sign conditions may not be preserved at the nodes of the scheme.

Theorem 2.14 [4] *In addition to the above assumptions, we assume that there exist $T > 0$, $\sigma > 0$ and $C_T > 0$ such that*

$$C_T E_{\phi_{\triangle x}}(0) \le \int_0^T \left(\|B_{\triangle x}^{1/2} \phi_{\triangle x}(t)\|_{\triangle x}^2 + (\triangle x)^\sigma \|V_{\triangle x}^{1/2} \phi_{\triangle x}(t)\|_{\triangle x}^2 \right) dt, \quad (2.22)$$

for every solution of $\phi'_{\triangle x}(t) + A_{\triangle x} \phi_{\triangle x}(t) = 0$ (uniform observability inequality with viscosity for the space semidiscretized linear conservative equation).

Then, the solutions of (2.21), with values in $X_{\triangle x}$, are well-defined on $[0, +\infty)$, and the energy of any solution satisfies

$$E_{u_{\triangle x}}(t) \lesssim T \max(\gamma_1, E_{u_{\triangle x}}(0)) L \left(\frac{1}{\psi^{-1}(\gamma_2 t)} \right),$$

for every $t \ge 0$, with $\gamma_1 \simeq \|B\|/\gamma_2$ and $\gamma_2 \simeq C_T/T(T^2\|B^{1/2}\|^4 + 1)$. Moreover, under (2.14), we have the simplified decay rate

$$E_{u_{\triangle x}}(t) \lesssim T \max(\gamma_1, E_{u_{\triangle x}}(0)) (H')^{-1} \left(\frac{\gamma_3}{t} \right),$$

for every $t \ge 0$, for some positive constant $\gamma_3 \simeq 1$.

The main assumption above is the uniform observability inequality (2.22), which is not easy to obtain in general, as already discussed. We stress again that, without the viscosity term, this uniform observability inequality does not hold true in general. As said previously, a typical example of a viscosity operator, for which uniform observability results do exist in the literature, is $V_{\triangle x} = \sqrt{A_{\triangle x}^* A_{\triangle x}}$.

2.3.3 Time Semidiscretizations

In this section, we show how to design a time semidiscrete numerical scheme for (2.13), while keeping uniform decay estimates.

As a first remark, let us note that a naive explicit discretization

$$\frac{u^{k+1} - u^k}{\Delta t} + Au^k + BF(u^k) = 0, \qquad t_k = k\Delta t,$$

is never suitable since

- it is instable and does never satisfy the basic CFL condition (note also that we are going to consider full discretizations in the next section),
- it does not preserve the energy of the conservative part.

Hence an implicit discretization is more appropriate.

Following [21], we choose an implicit mid-point rule:

$$\frac{u^{k+1} - u^k}{\Delta t} + A\left(\frac{u^k + u^{k+1}}{2}\right) + BF\left(\frac{u^k + u^{k+1}}{2}\right) = 0, \qquad t_k = k\Delta t.$$

Defining the energy by

$$E_{u^k} = \frac{1}{2}\|u^k\|_X^2,$$

we have

$$E_{u^{k+1}} - E_{u^k} = -\Delta t \left\langle B\left(\frac{u^k + u^{k+1}}{2}\right), F\left(\frac{u^k + u^{k+1}}{2}\right)\right\rangle_X.$$

The question is then to determine whether we have or not a uniform decay.

As before, the answer is **NO** in general, and in order to see that, we are first going to focus on the linear case.

2.3.3.1 Linear Case

Assuming that $F = \mathrm{id}_X$, we consider the equation

$$u' + Au + Bu = 0,$$

with the time semidiscrete model

$$\frac{u^{k+1} - u^k}{\Delta t} + A\left(\frac{u^k + u^{k+1}}{2}\right) + B\left(\frac{u^k + u^{k+1}}{2}\right) = 0.$$

We have

$$E_{u^k} = \frac{1}{2}\|u^k\|_X^2 \qquad \Rightarrow \qquad \frac{E_{u^{k+1}} - E_{u^k}}{\Delta t} = -\left\|B^{1/2}\left(\frac{u^k + u^{k+1}}{2}\right)\right\|_X^2 \leq 0,$$

and we investigate the question of knowing whether we have a uniform exponential decrease

$$E_{u^k} \leq C E_{u^0} \exp(-\delta k \Delta t).$$

A negative answer is given in [46] for a linear damped wave equation. We do not provide the details, which are similar to what has already been said.

Therefore, as before, filtering, or adding an appropriate viscosity term, is required in an attempt to recover uniformity.

In [21], the authors propose to consider the following general numerical scheme:

$$\begin{cases} \dfrac{\widetilde{u}^{k+1} - u^k}{\Delta t} + A\left(\dfrac{u^k + \widetilde{u}^{k+1}}{2}\right) + B\left(\dfrac{u^k + \widetilde{u}^{k+1}}{2}\right) = 0, \\[2ex] \dfrac{\widetilde{u}^{k+1} - u^{k+1}}{\Delta t} = \mathcal{V}_{\Delta t} u^{k+1}, \end{cases}$$

where $\mathcal{V}_{\Delta t} : X \to X$ ia s positive self-adjoint operator (viscosity operator). Written in an expansive way, we have

$$\frac{u^{k+1} - u^k}{\Delta t} + A\left(\frac{u^k + u^{k+1}}{2}\right) + B\left(\frac{u^k + u^{k+1}}{2}\right)$$
$$+ \frac{\Delta t}{2} B \mathcal{V}_{\Delta t} u^{k+1} + \mathcal{V}_{\Delta t} u^{k+1} + \frac{\Delta t}{2} A \mathcal{V}_{\Delta t} u^{k+1} = 0.$$

Considering the energy $E_{u^k} = \frac{1}{2}\|u^k\|_X^2$, we have

$$\frac{E_{u^{k+1}} - E_{u^k}}{\Delta t} = -\left\| B^{1/2}\left(\frac{u^k + \widetilde{u}^{k+1}}{2}\right) \right\|_X^2 - \|(\mathcal{V}_{\Delta t})^{1/2} u^{k+1}\|_X^2$$
$$- \frac{\Delta t}{2}\|\mathcal{V}_{\Delta t} u^{k+1}\|_X^2 \leq 0.$$

Now, the main (remarkable) result of [21] is the following.

Theorem 2.15 [21]

1. *We assume the uniform observability inequality with viscosity for the time semidiscretized linear conservative equation with viscosity:*

$$C_T \|\phi^0\|_X^2 \leq \Delta t \sum_{k=0}^{N-1} \left\| B^{1/2}\left(\frac{\phi^k + \widetilde{\phi}^{k+1}}{2}\right) \right\|_X^2$$
$$+ \Delta t \sum_{k=0}^{N-1} \|(\mathcal{V}_{\Delta t})^{1/2} \phi^{k+1}\|_X^2 + (\Delta t)^2 \sum_{k=0}^{N-1} \|\mathcal{V}_{\Delta t} \phi^{k+1}\|_X^2$$

for all solutions of

$$\begin{cases} \dfrac{\widetilde{\phi}^{k+1} - \phi^k}{\Delta t} + A\left(\dfrac{\phi^k + \widetilde{\phi}^{k+1}}{2}\right) = 0 \\ \dfrac{\widetilde{\phi}^{k+1} - \phi^{k+1}}{\Delta t} = V_{\Delta t}\phi^{k+1} \end{cases}.$$

Then the uniform exponential decrease holds, i.e.,

$$E_{u^k} \le CE_{u^0}\exp(-\delta k\Delta t),$$

for some positive constants C and δ not depending on the solution.
2. *The previous uniform observability inequality holds true if*

$$V_{\Delta t} = -(\Delta t)^2 A^2 = (\Delta t)^2 A^* A.$$

Note that this result provides a choice viscosity operator that always works, in the sense that the uniform observability inequality is systematically valid with this viscosity. Other choices are possible, such as

$$V_{\Delta t} = -(\mathrm{id}_X - (\Delta t)^2 A^2)^{-1}(\Delta t)^2 A^2.$$

The proof is done by resolvent estimates (Hautus test) and by carefully treating lower and higher frequencies.

2.3.3.2 Semilinear Case
Coming back to the general semilinear case, we consider the implicit midpoint time discretization of (2.13) given by

$$\begin{cases} \dfrac{\widetilde{u}^{k+1} - u^k}{\Delta t} + A\left(\dfrac{u^k + \widetilde{u}^{k+1}}{2}\right) + BF\left(\dfrac{u^k + \widetilde{u}^{k+1}}{2}\right) = 0, \\ \dfrac{\widetilde{u}^{k+1} - u^{k+1}}{\Delta t} = V_{\Delta t}u^{k+1}, \\ u^0 = u(0). \end{cases} \tag{2.23}$$

Written in an expansive way, (2.23) gives

$$\frac{u^{k+1} - u^k}{\Delta t} + A\left(\frac{u^k + u^{k+1}}{2}\right) + BF\left(\frac{u^k + (\mathrm{id}_X + \Delta t V_{\Delta t})u^{k+1}}{2}\right)$$
$$+ V_{\Delta t}u^{k+1} + \frac{\Delta t}{2}AV_{\Delta t}u^{k+1} = 0.$$

For the energy $E_{u^k} = \frac{1}{2}\|u^k\|_X^2$, we have

$$E_{u^{k+1}} - E_{u^k} = -\triangle t \left\langle B\left(\frac{u^k + \tilde{u}^{k+1}}{2}\right), F\left(\frac{u^k + \tilde{u}^{k+1}}{2}\right)\right\rangle_X$$

$$-\triangle t \,\|(\mathcal{V}_{\triangle t})^{1/2} u^{k+1}\|_X^2 - \frac{(\triangle t)^2}{2}\|\mathcal{V}_{\triangle t} u^{k+1}\|_X^2 \leq 0,$$

for every integer k.

Theorem 2.16 [4] *In addition to the above assumptions, we assume that there exist $T > 0$ and $C_T > 0$ such that, setting $N = [T/\triangle t]$ (integer part), we have*

$$C_T\|\phi^0\|_X^2 \leq \triangle t \sum_{k=0}^{N-1} \left\| B^{1/2}\left(\frac{\phi^k + \tilde{\phi}^{k+1}}{2}\right)\right\|_X^2$$

$$+\triangle t \sum_{k=0}^{N-1} \|(\mathcal{V}_{\triangle t})^{1/2}\phi^{k+1}\|_X^2 + (\triangle t)^2 \sum_{k=0}^{N-1} \|\mathcal{V}_{\triangle t}\phi^{k+1}\|_X^2,$$

for every solution of

$$\begin{cases} \dfrac{\tilde{\phi}^{k+1} - \phi^k}{\triangle t} + A\left(\dfrac{\phi^k + \tilde{\phi}^{k+1}}{2}\right) = 0, \\[2mm] \dfrac{\tilde{\phi}^{k+1} - \phi^{k+1}}{\triangle t} = \mathcal{V}_{\triangle t}\phi^{k+1}, \end{cases}$$

(uniform observability inequality with viscosity for the time semidiscretized linear conservative equation with viscosity).

Then, the solutions of (2.23) are well-defined on $[0, +\infty)$ and, the energy of any solution satisfies

$$E_{u^k} \lesssim T\max(\gamma_1, E_{u^0})\, L\left(\frac{1}{\psi^{-1}(\gamma_2 k\triangle t)}\right),$$

for every integer k, with $\gamma_2 \simeq C_T/T(1 + e^{2T\|B\|}\max(1, T\|B\|))$ and $\gamma_1 \simeq \|B\|/\gamma_2$. Moreover, under (2.14), we have the simplified decay rate

$$E_{u^k} \lesssim T\max(\gamma_1, E_{u^0})\,(H')^{-1}\left(\frac{\gamma_3}{k\triangle t}\right),$$

for every integer k, for some positive constant $\gamma_3 \simeq 1$.

2.3.4 Full Discretizations

Results for full discretization schemes can be obtained in an automatic way, from the previous time discretization and space discretization results: indeed, following [21], we note that the results for time semidiscrete approximation schemes are actually valid for a class of abstract systems depending on a parameter, uniformly with respect to this parameter that is typically the space mesh parameter $\triangle x$. Then, using the results obtained for space semidiscretizations, we infer the desired uniform properties for fully discrete schemes. In other words, we first discretize in space and then in time. We refer to [4] for the detailed definition of that abstract class.

Theorem 2.17 [4] *Under the assumptions of Theorems 2.6, 2.14, and 2.16, the solutions of*

$$
\begin{cases}
\dfrac{\widetilde{u}_{\triangle x}^{k+1} - u_{\triangle x}^{k}}{\triangle t} + A_{\triangle x}\left(\dfrac{u_{\triangle x}^{k} + \widetilde{u}_{\triangle x}^{k+1}}{2}\right) + B_{\triangle x}F_{\triangle x}\left(\dfrac{u_{\triangle x}^{k} + \widetilde{u}_{\triangle x}^{k+1}}{2}\right) \\[2ex]
\quad + \mathcal{V}_{\triangle x}\left(\dfrac{u_{\triangle x}^{k} + \widetilde{u}_{\triangle x}^{k+1}}{2}\right) = 0, \\[2ex]
\dfrac{\widetilde{u}_{\triangle x}^{k+1} - u_{\triangle x}^{k+1}}{\triangle t} = \mathcal{V}_{\triangle t}u_{\triangle x}^{k+1},
\end{cases}
$$

are well-defined for every integer k, for every initial condition $u_{\triangle x}^{0} \in X_{\triangle x}$, and for every $\triangle x \in (0, \triangle x_0)$, and the energy of any solution satisfies

$$
\frac{1}{2}\|u_{\triangle x}^{k}\|_{\triangle x}^{2} = E_{u_{\triangle x}^{k}} \le T\max(\gamma_1, E_{u_{\triangle x}^{0}}) L\left(\frac{1}{\psi^{-1}(\gamma_2 k\triangle t)}\right),
$$

for every integer k, with $\gamma_2 \simeq C_T/T(1 + e^{2T\|B\|}\max(1, T\|B\|))$ and $\gamma_1 \simeq \|B\|/\gamma_2$. Moreover, under (2.14), we have the simplified decay rate

$$
E_{u_{\triangle x}^{k}} \le T\max(\gamma_1, E_{u_{\triangle x}^{0}}) (H')^{-1}\left(\frac{\gamma_3}{k\triangle t}\right),
$$

for every integer k, for some positive constant $\gamma_3 \simeq 1$.

As an example, we consider the nonlinear damped wave equation

$$
\partial_{tt}u(t, x) - \triangle u(t, x) + b(x)\rho(x, \partial_t u(t, x)) = 0,
$$

with appropriate assumptions on ρ, as discussed previously. We first semidiscretize in space with finite differences with the viscosity operator $\mathcal{V}_{\triangle x} = -(\triangle x)^2\triangle_{\triangle x}$, obtaining

$$
u''_{\triangle x,\sigma}(t) - \triangle_{\triangle x}u_{\triangle x,\sigma}(t) + b_{\triangle x}(x_\sigma)\rho(x_\sigma, u'_{\triangle x,\sigma}(t)) - (\triangle x)^2\triangle_{\triangle x}u'_{\triangle x}(t) = 0,
$$

where σ is an index for the mesh. The solutions of that system have the uniform energy decay rate $L(1/\psi^{-1}(t))$ (up to some constants).

We next discretize in time, obtaining

$$
\begin{cases}
\dfrac{\widetilde{u}^{k+1}_{\Delta x,\sigma} - u^k_{\Delta x,\sigma}}{\Delta t} = \dfrac{v^k_{\Delta x,\sigma} + \widetilde{v}^{k+1}_{\Delta x,\sigma}}{2}, \\[2ex]
\dfrac{\widetilde{v}^{k+1}_{\Delta x,\sigma} - v^k_{\Delta x,\sigma}}{\Delta t} - \triangle_{\Delta x,\sigma}\dfrac{u^k_{\Delta x,\sigma}+\widetilde{u}^{k+1}_{\Delta x,\sigma}}{2} + b_{\Delta x,\sigma}(x_\sigma)\rho\left(x_\sigma,\dfrac{v^k_{\Delta x,\sigma}+\widetilde{v}^{k+1}_{\Delta x,\sigma}}{2}\right) \\[2ex]
\quad - (\Delta x)^2 \triangle_{\Delta x,\sigma}\dfrac{v^k_{\Delta x,\sigma}+\widetilde{v}^{k+1}_{\Delta x,\sigma}}{2} = 0, \\[2ex]
\dfrac{\widetilde{u}^{k+1}_{\Delta x,\sigma} - u^{k+1}_{\Delta x,\sigma}}{\Delta t} = -(\Delta t)^2 \triangle^2_{\Delta x,\sigma} u^{k+1}_{\Delta x,\sigma}, \\[2ex]
\dfrac{\widetilde{v}^{k+1}_{\Delta x,\sigma} - v^{k+1}_{\Delta x,\sigma}}{\Delta t} = -(\Delta t)^2 \triangle^2_{\Delta x,\sigma} v^{k+1}_{\Delta x,\sigma},
\end{cases}
$$

and according to Theorem 2.17, the solutions of that system have the uniform energy decay rate $L(1/\psi^{-1}(t))$ (up to some constants).

2.3.5 Conclusion and Open Problems

2.3.5.1 (Un)-Boundedness of B

We have assumed B bounded: this involves the case of local or nonlocal internal dampings, but this does not cover, for instance, the case of boundary dampings.

For example, let us consider the linear 1D wave equation with boundary damping

$$
\begin{cases}
\partial_{tt}u - \partial_{xx}u = 0, & t \in (0,+\infty),\ x \in (0,1), \\
u(t,0) = 0,\quad \partial_x u(t,1) = -\alpha\partial_t u(t,1), & t \in (0,+\infty),
\end{cases}
$$

with $\alpha > 0$. The energy

$$
E(t) = \frac{1}{2}\int_0^1 \left(|\partial_t u(t,x)|^2 + |\partial_x u(t,x)|^2\right) dx
$$

decays exponentially. It is proved in [41] that the solutions of

$$
\begin{cases}
u''_{\Delta x} - \triangle_{\Delta x}u_{\Delta x} - (\Delta x)^2 \triangle_{\Delta x}u'_{\Delta x} = 0, \\
u_{\Delta x}(t,0) = 0,\quad D_{\Delta x}u_{\Delta x}(t,1) = -\alpha\partial_t u(t,1),
\end{cases}
$$

(regular finite-difference space semidiscrete model with viscosity) with

$$D_{\triangle x} = \frac{1}{\triangle x} \begin{pmatrix} -1 & 1 & \dots & 0 \\ 0 & \ddots & \ddots & \vdots \\ \vdots & \ddots & \ddots & 1 \\ 0 & \dots & 0 & -1 \end{pmatrix}$$

(forward finite-difference operator), have a *uniform exponential decay*. But this fails without the viscosity term.

This example is not covered by our results.

We also mention [6] where a sufficient condition (based on energy considerations and not on viscosities) ensuring uniformity is provided and is applied to 1D mixed finite elements and to polynomial Galerkin approximations in hypercubes. The extension to the semilinear setting is open.

2.3.5.2 More General Nonlinear Models
If the nonlinearity F involves an unbounded operator, for example

$$u'(t) + Au(t) + BF(u(t), \nabla u(t)) = 0,$$

then the situation is widely open.

For instance, one can think of investigating semilinear wave equations with strong damping

$$\partial_{tt}u - \triangle u - a\triangle\partial_t u + b\partial_t u + g \star \triangle u + f(u) = 0,$$

with f not too much superlinear. There are many possible variants, with boundary damping, with delay, with nonlocal terms (convolution), etc. Many results do exist in the continuous setting, but the investigation of whether the uniform decay is preserved under discretization or not remains to be done.

2.3.5.3 Energy with Nonlinear Terms
Another case which is not covered by our results is the case of semilinear wave equations with locally distributed damping

$$\partial_{tt}u - \triangle u + a(x)\partial_t u + f(u) = 0,$$

with Dirichlet boundary conditions, a a nonnegative bounded function which is positive on an open subset ω of Ω, and f a function of class C^1 such that $f(0) = 0$, $sf(s) \geq 0$ for every $s \in \mathbb{R}$ (defocusing case), $|f'(s)| \leq C|s|^{p-1}$ with $p \leq n/(n-2)$ (energy subcritical). It is proved in [47] (see also [17, 25]) that, under geometric conditions on ω, the energy involving a nonlinear term

$$\int_\Omega \left((\partial_t u)^2 + \|\nabla u\|^2 + F(u)\right) dx,$$

with $F(s) = \int_0^s f$, decays exponentially in time along any solution. The investigation of the uniform decay for approximation schemes seems to be open. In particular, it would be interesting to know whether microlocal arguments withstand the discretization procedure. In particular we raise the following informal question:

Do microlocalization and discretization commute?

2.3.5.4 Uniform Polynomial Energy Decay without Observability

Let us assume that $F = \mathrm{id}_X$ in (2.13) (linear case). As proved in Theorem 2.3, observability for the conservative linear equation is equivalent to the exponential decay for the linear damped equation.

If observability fails for the conservative equation, then we do not have an exponential decay, however we can still hope to have, for instance, a polynomial decay. This is the case for some weakly damped wave equations when GCC fails (see [9]).

In such cases, it is interesting to investigate the question of proving a uniform polynomial decay rate for space and/or time semidiscrete approximations. In [1], such results are stated for second-order linear equations, with appropriate viscosity terms, and under adequate spectral gap conditions. The extension to a more general framework (weaker assumptions, full discretizations), and to semilinear equations, seems to be open.

References

[1] F. Abdallah, S. Nicaise, J. Valein, A. Wehbe, Uniformly exponentially or polynomially stable approximations for second order evolution equations and some applications, *ESAIM Control Optim. Calc. Var.* **19** (2013), no. 3, 844–87.

[2] F. Alabau-Boussouira, Convexity and weighted integral inequalities for energy decay rates of nonlinear dissipative hyperbolic systems, *Appl. Math. Optim.* **51** (2005), 61–105.

[3] F. Alabau-Boussouira, K. Ammari, Sharp energy estimates for nonlinearly locally damped PDE's via observability for the associated undamped system, *J. Funct. Anal.* **260** (2011), 2424–50.

[4] F. Alabau-Boussouira, Y. Privat, E. Trélat, Nonlinear damped partial differential equations and their uniform discretizations, Preprint Hal/Arxiv (2015).

[5] K. Ammari, M. Tucsnak, Stabilization of second order evolution equations by a class of unbounded operators, *ESAIM: Cont. Optim. Calc. Var.* **6** (2001), 361–86.

[6] H.T. Banks, K. Ito, C. Wang, Exponentially stable approximations of weakly damped wave equations. Estimation and control of distributed parameter systems (Vorau, 1990), 1–33, *Internat. Ser. Numer. Math.*, 100, Birkhäuser, Basel, 1991.

[7] H.T. Banks, K. Ito, Approximation in LQR problems for infinite dimensional systems with unbounded input operators, *J. Math. Systems Estim. Control* **7** (1997), no. 1, 1–34.

[8] H.T. Banks, K. Kunisch, The linear regulator problem for parabolic systems, *SIAM J. Control Optim.* **22** (1984), no. 5, 684–98.

[9] C. Bardos, G. Lebeau, J. Rauch, Sharp sufficient conditions for the observation, control, and stabilization of waves from the boundary, *SIAM J. Control Optim.* **30** (1992), no. 5, 1024–65.

[10] T. Cazenave, A. Haraux, An introduction to semilinear evolution equations, Translated from the 1990 French original by Yvan Martel and revised by the authors. *Oxford Lecture Series in Mathematics and Its Applications*, 13. The Clarendon Press, Oxford University Press, New York, 1998.

[11] J.-M. Coron, Control and nonlinearity, *Mathematical Surveys and Monographs*, 136, American Mathematical Society, Providence, RI, 2007.

[12] J.-M. Coron, E. Trélat, Global steady-state controllability of 1-D semilinear heat equations, *SIAM J. Control Optim.* **43** (2004), no. 2, 549–69.

[13] J.-M. Coron, E. Trélat, Global steady-state stabilization and controllability of 1-D semilinear wave equations, *Commun. Contemp. Math.* **8** (2006), no. 4, 535–67.

[14] J.-M. Coron, E. Trélat, R. Vazquez, Control for fast and stable laminar-to-high-Reynolds-numbers transfer in a 2D Navier–Stokes channel flow, *Discrete Contin. Dyn. Syst. Ser. B* **10** (2008), no. 4, 925–56.

[15] R.F. Curtain, H. Zwart, An introduction to infinite-dimensional linear systems theory, *Texts in Applied Mathematics*, 21, Springer-Verlag, New York, 1995.

[16] C.M. Dafermos, Contraction semi-groups and trend to equilibrium in continuum mechanics, *Lecture Notes in Mathematics*, Vol. 503, pp. 295–306, Springer-Verlag, New York/Berlin, 1976.

[17] B. Dehman, G. Lebeau, E. Zuazua, Stabilization and control for the subcritical semilinear wave equation, *Ann. Sci. Ecole Norm. Sup.* **36** (2003), no. 4, 525–51.

[18] K.-J. Engel, R. Nagel, One-parameter semigroups for linear evolution equations, *Graduate Texts in Mathematics*, 194, Springer-Verlag, New York, 2000.

[19] S. Ervedoza, Spectral conditions for admissibility and observability of wave systems: Applications to finite element schemes, *Numer. Math.* **113** (2009), no. 3, 377–415.

[20] S. Ervedoza, Observability properties of a semi-discrete 1D wave equation derived from a mixed finite element method on nonuniform meshes, *ESAIM Control Optim. Calc. Var.* **16** (2010), no. 2, 298–326.

[21] S. Ervedoza, E. Zuazua, Uniformly exponentially stable approximations for a class of damped systems, *J. Math. Pures Appl.* **91** (2009), 20–48.

[22] J.S. Gibson, The Riccati integral equations for optimal control problems on Hilbert spaces, *SIAM J. Control Optim.* **17** (1979), 537–65.

[23] A. Haraux, Une remarque sur la stabilisation de certains systèmes du deuxième ordre en temps, *Portugal. Math.* **46** (1989), no. 3, 245–58.

[24] J.A. Infante, E. Zuazua, Boundary observability for the space semi-discretizations of the 1D wave equation, *M2AN Math. Model. Numer. Anal.* **33** (1999), no. 2, 407–38.

[25] R. Joly, C. Laurent, Stabilization for the semilinear wave equation with geometric control condition, *Anal. PDE* **6** (2013), no. 5, 1089–119.

[26] F. Kappel, D. Salamon, An approximation theorem for the algebraic Riccati equation, *SIAM J. Control Optim.* **28** (1990), no. 5, 1136–47.

[27] V. Komornik, Rapid boundary stabilization of linear distributed systems, *SIAM J. Control Optim.* **35** (1997), no. 5, 1591–613.

[28] S. Labbé, E. Trélat, Uniform controllability of semidiscrete approximations of parabolic control systems, *Syst. Cont. Lett.* **55** (2006), no. 7, 597–609.

[29] I. Lasiecka, R. Triggiani, Control theory for partial differential equations: continuous and approximation theories. I. Abstract parabolic systems, *Encyclopedia of Mathematics and Its Applications*, 74, Cambridge University Press, Cambridge, 2000.

[30] I. Lasiecka, R. Triggiani, Control theory for partial differential equations: continuous and approximation theories. II. Abstract hyperbolic-like systems over a finite time horizon, *Encyclopedia of Mathematics and Its Applications*, 75, Cambridge University Press, Cambridge, 2000.

[31] K. Liu, Locally distributed control and damping for the conservative systems, *SIAM J. Control Optim.* **35** (1997), no. 5, 1574–90.

[32] Z. Liu, S. Zheng, *Semigroups Associated with Dissipative Systems*, Chapman & Hall/CRC Research Notes in Mathematics, 398, Boca Raton, FL, 1999.

[33] A. Marica, E. Zuazua, Localized solutions for the finite difference semi-discretization of the wave equation, *C. R. Math. Acad. Sci. Paris* **348** (2010), no. 11–12, 647–52.

[34] L. Miller, Resolvent conditions for the control of unitary groups and their approximations, *J. Spectr. Theory* **2** (2012), no. 1, 1–55.

[35] A. Münch, A.F. Pazoto, Uniform stabilization of a viscous numerical approximation for a locally damped wave equation, *ESAIM Control Optim. Calc. Var.* **13** (2007), no. 2, 265–93.

[36] K. Ramdani, T. Takahashi, M. Tucsnak, Internal stabilization of the plate equation in a square: The continuous and the semi-discretized problems, *J. Math. Pures Appl.* **85** (2006), 17–37.

[37] K. Ramdani, T. Takahashi, M. Tucsnak, Uniformly exponentially stable approximations for a class of second order evolution equations, *ESAIM: Control Optim. Calc. Var.*, **13** (2007), no. 3, 503–27.

[38] D.L. Russell, Controllability and stabilizability theory for linear partial differential equations: Recent progress and open questions, *SIAM Rev.* **20** (1978), no. 4, 639–739.

[39] M. Schmidt, E. Trélat, Controllability of Couette flows, *Commun. Pure Appl. Anal.* **5** (2006), no. 1, 201–11.

[40] L.R. Tcheugoué Tébou, E. Zuazua, Uniform exponential long time decay for the space semi discretization of a locally damped wave equation via an artificial numerical viscosity, *Numer. Math.* **95** (2003), no. 3, 563–98.

[41] L.R. Tcheugoué Tébou, E. Zuazua, Uniform boundary stabilization of the finite difference space discretization of the 1-d wave equation, *Adv. Comput. Math.* **26** (2007), 337–65.

[42] E. Trélat, *Contrôle Optimal: Théorie & Applications*, Vuibert, Collection "Mathématiques Concrètes" (2005), 246 pages.

[43] E. Trélat, *Control in Finite and Infinite Dimension*, BCAM Springer Briefs in Mathematics, to appear.

[44] M. Tucsnak, G. Weiss, Observation and control for operator semigroups, *Birkhäuser Advanced Texts: Basler Lehrbücher*, Birkhäuser Verlag, Basel, 2009.

[45] J. Zabczyk, Mathematical control theory: An introduction, *Systems & Control: Foundations & Applications*, Birkhäuser Boston, Inc., Boston, MA, 1992.

[46] X. Zhang, C. Zheng, E. Zuazua, Time discrete wave equations: boundary observability and control, *Discrete Cont. Dynam. Syst.* **23** (2009), no. 1–2, 571–604.

[47] E. Zuazua, Exponential decay for the semilinear wave equation with locally distributed damping, *Comm. Partial Diff. Eq.* **15** (1990), no. 2, 205–35.

[48] E. Zuazua, Propagation, observation, and control of waves approximated by finite difference methods, *SIAM Rev.* **47** (2005), no. 2, 197–243.

SORBONNE UNIVERSITÉS, UPMC UNIV PARIS 06, CNRS UMR 7598, LABORATOIRE JACQUES-LOUIS LIONS, INSTITUT UNIVERSITAIRE DE FRANCE, F-75005, PARIS, FRANCE
E-mail address: emmanuel.trelat@upmc.fr

3

A Null-Controllability Result for the Linear System of Thermoelastic Plates with a Single Control

CARLOS CASTRO AND LUZ DE TERESA

Abstract

We present a null-controllability result for the linear system of thermoelastic plates when we consider a single control acting in the parabolic equation. This is an extension of our previous work [Castro and de Teresa, *J. Math. Anal. App.* 428 (2) (2015), 772–93] where we proved a similar result when considering two different controls acting in each one of the equations of the system.

Mathematics Subject Classification 2010. 93B05, 93C20, 35B37
Key words and phrases. Stabilization, observability, discretized control problem

Contents

3.1 Introduction and Main Results

In this chapter we present a null-controllability result for the linear system of thermoelastic plates. Let Ω be a bounded, open, connected set in $\mathbb{R}^2$ with boundary $\partial\Omega$ of class C^3, $T > 0$ and $Q = \Omega \times (0, T)$ and $\Sigma = \partial\Omega \times (0, T)$. We consider the following controlled system which describes the small vibrations of a thin, homogeneous, isotropic, hinged thermoelastic plate:

$$\begin{cases} w_{tt} + \gamma\Delta w_{tt} + \Delta^2 w + \beta_1\Delta\theta = 0 & \text{in } Q \\ \beta_2\theta_t - \Delta\theta - \varepsilon\Delta w_t = f_p & \text{in } Q \\ w = \Delta w = 0 & \text{on } \Sigma \\ \theta = 0 & \text{on } \Sigma \\ w(0) = w^0; \ w_t(0) = w^1; \ \theta(0) = \theta^0, & \text{in } \Omega. \end{cases} \qquad (3.1)$$

Here $\varepsilon > 0$, $\gamma > 0$, $\beta_i > 0$ (for $i = 1, 2$) are constants. The function w represents the deflection of the middle plane of the plate from its equilibrium position (in fact Ω), while θ is the temperature at this middle plane. The coupling term $\varepsilon\Delta w_t$ takes into account the heat induced by the high frequency vibrations of the plate, and $f_p(x, t)$ is the control acting on a suitable proper subset of Ω and it represents a local heat source. We refer to [9] for a heuristic derivation of the model.

The controllability of this model was previously studied in [5] when considering two different controls f_h and f_p, acting in the hyperbolic and parabolic equations, respectively. In this chapter we revisit the approach introduced there to see how one can obtain a similar result with one single control acting in the parabolic equation.

Control models related to (3.1) have been extensively studied in the last two decades. In the case $\gamma = 0$ we refer to [1–3, 17, 20]. The case $\gamma > 0$ was treated in [1, 16] where a single control is considered, but acting in the whole domain.

There are also interesting results related to the reachability of the thermoelastic system in [9], where other boundary conditions are considered. Also partial results, with the plates being exactly controlled and the heat approximately controlled, can be found in [7, 20]. It is also worth mentioning the results in [8] for thermoelastic beams.

Our main contribution is the null-controllability when $\varepsilon > 0$ is small enough, with a single control supported in an open subset of Ω and acting on the parabolic equation. The control must be supported on a neighborhood of a sufficiently large part of the boundary. The proof follows the approach introduced in our previous work [5] where we considered two controls. The result is deduced from the controllability of a reduced system where the coupling is present only in the plate model, and we refer to that as the cascade system. The coupling term in the heat equation is treated as a small perturbation of this cascade system. In this chapter we give an alternative proof for the controllability of this cascade system which provides an explicit formula for both the control acting in the parabolic and hyperbolic equations. This allows us to eliminate the control acting in the hyperbolic one.

To describe the result in a more precise way we first introduce the spaces of existence of solutions and where the controls are posed. To this end, we define H_D^α, $\alpha \in \mathbb{N}$, the usual scale of spaces associated with the Laplacian operator in Ω with Dirichlet boundary conditions:

$$H_D^0(\Omega) = L^2(\Omega), \quad H_D^1(\Omega) = H_0^1(\Omega), \quad H_D^2(\Omega) = H^2(\Omega) \cap H_0^1(\Omega),$$

$$H_D^3(\Omega) = \{u \in H^3(\Omega)|u = \Delta u = 0 \text{ on } \partial\Omega\}.$$

We denote their dual spaces as

$$(H_D^\alpha)'(\Omega) = H_D^{-\alpha}(\Omega).$$

For brevity we will write H_D^α and $H_D^{-\alpha}$. We also define

$$V' = H_D^{-2} \times H_D^0 \times H_D^0.$$

We introduce now some sets and functions related to the control of the hyperbolic equation. Given $x_0 \in \mathbb{R}^2$ we set

$$m(x) = x - x_0; \quad R(x_0) = \max_{x \in \Omega} |x - x_0|;$$

$$M(x_0) = \{x \in \partial\Omega | m(x) \cdot \nu > 0\}, \quad \Sigma(x_0) = M(x_0) \times (0, T).$$

We assume that $\omega \subset \Omega$ is an open neighborhood of a set $M(x_0)$, for some $x_0 \in \mathbb{R}^2$.

The control problem we treat is the following: Given $(w^0, w^1, \theta^0) \in V'$, $T > 0$, and $\omega \subset \Omega$ are an open nonempty subset, find the control f_p, supported in ω such that the solution of (3.1) satisfies

$$(w(T), w_t(T), \theta(T)) = (0, 0, 0). \tag{3.2}$$

Our main result is the following:

Theorem 3.1 *Let $\omega \subset \Omega$ be a neighborhood of the set $M(x_0)$ for some x_0 in $\mathbb{R}$ and define $T_0 = 2\sqrt{\gamma}\max_{x \in \Omega}|m(x)|$. There exists $\varepsilon_0 > 0$ such that for every $\varepsilon_0 > \varepsilon \geq 0$ and any $T > T_0$ the following null-controllability result holds: Given any initial data $(w, w_t, \theta) \in V'$, there exists a control $f_p \in L^2(0, T; H_D^{-2}(\Omega))$, supported in ω, such that the corresponding solution to (3.1) satisfies (3.2).*

The rest of this chapter is divided as follows: In Section 3.2 we recall some results in [5], in particular the well-posedness of system (3.1) for the initial data and controls in the functional spaces stated in Theorem 3.1, and the fact that the solutions of (3.1) can be written in series form in such a way that each term satisfies a new system of equations, similar to (3.1), but coupled in cascade form. In this way, the proof of Theorem 3.1 is reduced to a controllability result for the associated cascade systems. In Section 3.3 we prove this null controllability for the cascade systems introduced in the previous section. Finally, in the appendix we prove some technical results needed in this chapter.

3.2 Solutions in Series Form

The results in this section are proved in [5]:

Theorem 3.2 *Given* $(\Delta w_0, w_1, \theta_0) \in \mathcal{H}$ *and* $f_p \in L^1(0, T; H_D^{-2})$ *there exists a unique solution of system (3.1) in the space*

$$w \in C([0, T]; H_D^0), \quad w_t \in C([0, T]; H_D^{-1}), \quad \theta \in C([0, T]; H_D^{-2}).$$

In this section we show that, under certain conditions, the solutions of (3.1) can be written in series form where each term of the series solves a suitable system of equations that can be solved in cascade.

We assume that $\varepsilon > 0$ is small and that we can decompose the solution and controls in (3.1) as a power series in ε. That is,

$$w = \sum_{k=0}^{\infty} \varepsilon^k w^k, \quad \theta = \sum_{k=0}^{\infty} \varepsilon^k \theta^k, \quad f_p = \sum_{k=0}^{\infty} \varepsilon^k f_p^k. \tag{3.3}$$

Then, if we substitute these expressions in the system (3.1) and make equal the terms with the same power in ε we obtain that w^0, θ^0, f_p^0 must solve

$$\begin{cases} w_{tt}^0 - \gamma \Delta w_{tt}^0 + \Delta^2 w^0 + \beta_1 \Delta \theta^0 = 0 & \text{in } Q \\ \beta_2 \theta_t^0 - \Delta \theta^0 = f_p^0 & \text{in } Q \\ w^0 = \Delta w^0 = 0 & \text{on } \Sigma \\ \theta^0 = 0 & \text{on } \Sigma \\ w^0(0) = w_0; \ w_t^0(0) = w_1; \ \theta^0(0) = \theta_0, & \text{in } \Omega, \end{cases} \tag{3.4}$$

and for $k \geq 1$, w^k, θ^k, f_p^k is the solution of

$$
\begin{cases}
w_{tt}^k - \gamma \Delta w_{tt}^k + \Delta^2 w^k + \beta_1 \Delta \theta^k = 0 & \text{in } Q \\
\beta_2 \theta_t^k - \Delta \theta^k - \Delta w_t^{k-1} = f_p^k & \text{in } Q \\
w^k = \Delta w^k = 0 & \text{on } \Sigma \\
\theta^k = 0 & \text{on } \Sigma \\
w^k(0) = 0; \ w_t^k(0) = 0; \ \theta^k(0) = 0, & \text{in } \Omega.
\end{cases}
\tag{3.5}
$$

Proposition 3.3 Let $y_0 = (\Delta w_0, w_1, \theta_0) \in \mathcal{H}$ and $f_p^k \in L^1(0, T; H_D^{-2})$ for $k \geq 0$. Then, there exists a unique solution of system (3.4) and systems (3.5) in the space

$$
w^k \in C([0, T]; H_D^0), \quad w_t^k \in C([0, T]; H_D^{-1}), \quad \theta^k \in C([0, T]; H_D^{-2}). \tag{3.6}
$$

Moreover, if

$$
\|f_p^k\|_{L^2(0,T;H_D^{-2})} \leq C \|\Delta w_t^{k-1}\|_{L^2(0,T;H_D^{-3})} \tag{3.7}
$$

for some constant $C > 0$ independent of $k \geq 1$, then there exists ε_0 such that for $0 < \varepsilon < \varepsilon_0$ we have

$$
f_p = \sum_{k=0}^{\infty} \varepsilon^k f_p^k \in L^2(0, T; H_D^{-2}), \tag{3.8}
$$

and the unique mild solution (w, θ) of (3.1) satisfies

$$
w = \sum_{k=0}^{\infty} \varepsilon^k w^k \in C([0, T]; H_D^0), \quad w_t = \sum_{k=0}^{\infty} \varepsilon^k w_t^k \in C([0, T]; H_D^{-1}),
$$
$$
\theta = \sum_{k=0}^{\infty} \varepsilon^k \theta^k \in C([0, T]; H_D^{-2}). \tag{3.9}
$$

3.3 Control of the Cascade System

In this section we prove that systems (3.4) and (3.5) are null-controllable at time $T > 0$ and that the associate controls satisfy the bounds given in (3.7). Both situations can be reduced to study the null-controllability properties of the general system

$$
\begin{cases}
v_{tt} - \gamma \Delta v + \Delta^2 v + \beta_1 \Delta \eta = f_h & \text{in } Q \\
\beta_2 \eta_t - \Delta \eta = G + f_p & \text{in } Q \\
v = \Delta v = 0 & \text{on } \Sigma \\
\eta = 0 & \text{on } \Sigma \\
v(0) = v_0; \ v_t(0) = v_1; \ \eta(0) = \eta_0, & \text{in } \Omega,
\end{cases}
\tag{3.10}
$$

where the function $G(x, t)$ and initial data (v_0, v_1, η_0) are given, and (f_h, f_p) are two controls acting in the hyperbolic and parabolic equations, respectively. As we said in the introduction our result is stated for a single control f_p acting on the parabolic equation. However, it is convenient to write the system with two controls since it is used in the proof. The analogy to the existence result stated in Theorem 3.2 for system (3.10) is also proved in [5].

In order to state the result we introduce the weight $e_M(t) = \exp(M/(T-t))$ and define the Hilbert space

$$L^2(e_M, H_D^{-3}) = \left\{ G \in L^2(0, T; H_D^{-3}(\Omega)) \text{ s.t. } \int_0^T e_M(t) \|G(t)\|_{H_D^{-3}}^2 < \infty \right\}.$$

The main result of this section is the following:

Proposition 3.4 Let ω be a neighborhood of $M(x_0)$ and $T > 2\sqrt{\gamma}$ $\max_{x \in \Omega} |m(x)|$. Then there exists $M' > 0$ such that for every $G \in L^2(e_M, H_D^{-3})$ with $M > M'$ and every $(v_0, v_1, \eta_0) \in V' = H_D^0 \times H_D^{-1} \times H_D^{-2}$ there exists a control $f_p \in L^2(0, T; H_D^{-2})$ with supp $f_p \subset ([0, T] \times \omega)$ such that the corresponding solution to (3.10), with $f_h = 0$, satisfies

$$(v(T), v_t(T), \eta(T)) = (0, 0, 0)$$

and

$$\|f_p\|_{L^2(0, T; H_D^{-2})} \leq C(\|(v_0, v_1, \eta_0)\|_{V'} + \|G\|_{L^2(0, T; H_D^{-3})}). \qquad (3.11)$$

Moreover,

$$\Delta \eta_t \in L^2(e_M, H_D^{-3}). \qquad (3.12)$$

The proof of Theorem 3.1 is a direct consequence of Proposition 3.3 and the estimates given by Proposition 3.4 when applied to systems (3.4) and (3.5). In the rest of this section we prove Proposition 3.4.

To prove this result we adopt the variational approach in [4] for the heat equation, based on ideas in [7]. This approach is different from the one used in our previous chapter [5] where we adopted the HUM method. In this way, we are able to construct controls and solutions that decay exponentially in time, as $t \to T$. This property is used to check that $\Delta w_t^{k-1} \in L^2(e_M, H_D^{-3})$ for each $k \geq 1$ in (3.5) and therefore all these systems satisfy the hypothesis of Proposition 3.4.

The following observability results for the plate and heat equations are needed.

Lemma 3.5 (Observability for the plate model) Suppose that ω_h is a neighborhood of a set $M(x_0)$ for some x_0 in $\mathbb{R}$ and $T > 2\sqrt{\gamma}R(x_0)$. Then, there exists $C > 0$ such that

$$\mathcal{E}(0) + \int_0^T \int_\Omega e^{\frac{-M_0}{T-t}} |\nabla \Delta \varphi(x,t)|^2 dx$$

$$\leq C \left(\int_0^T \int_{\omega_h} e^{\frac{-M_0}{T-t}} |\nabla \Delta \varphi|^2 + \int_0^T \int_\Omega e^{\frac{-M_0}{T-t}} |\Gamma \varphi_{tt} + \Delta^2 \varphi|^2 \right) \quad (3.13)$$

for any $M_0 > 0$, $t \in [0,T]$, and any

$$\varphi \in C([0,T]; H_D^3) \cap C^1([0,T]; H_D^2),$$

where

$$\mathcal{E}(t) = \frac{1}{2} \int_\Omega (|\nabla \varphi_t(x,t)|^2 + \gamma |\Delta \varphi_t(x,t)|^2 + |\nabla \Delta \varphi(x,t)|^2) dx. \quad (3.14)$$

Lemma 3.6 (Observability for the heat equation) Suppose that $\omega_p \subset \Omega$ is a nonempty open subset. Then, there exist $C > 0$ and $M_0 > 0$, such that the following observability inequality holds,

$$\int_\Omega |\Delta \psi(0)|^2 + \int_0^T \int_\Omega e^{\frac{-M_0}{T-t}} |\nabla \Delta \psi|^2$$

$$\leq C \left(\int_0^T \int_{\omega_p} e^{\frac{-M_0}{T-t}} |\Delta \psi|^2 + \int_0^T \int_\Omega e^{\frac{-M_0}{T-t}} |-\nabla \psi_t - \nabla \Delta \psi|^2 \right). \quad (3.15)$$

Lemma 3.5 is proved in the appendix below, while Lemma 3.6 is stated in [5] as a consequence of a well-known Carleman inequality for the heat equation. As a direct consequence of the above lemmas we have the following result for the cascade system:

Proposition 3.7 (Observability for the coupled system) Suppose that ω_h is a neighborhood of $M(x_0)$, $T > 2\sqrt{\gamma}R(x_0)$, and that $\omega_p \subset \Omega$ is a nonempty open subset. Then, there exist $C > 0$ and $M_0 > 0$, such that the following observability inequality holds,

$$\int_\Omega |\Delta \psi(0)|^2 + \int_0^T \int_\Omega e^{\frac{-M_0}{T-t}} |\nabla \Delta \psi|^2 + \int_\Omega |\nabla \varphi_t(0)|^2 + \gamma \int_\Omega |\Delta \varphi_t(0)|^2$$

$$+ \int_\Omega |\nabla \Delta \varphi(0)|^2 \leq C \left(\int_0^T \int_{\omega_h} e^{\frac{-M_0}{T-t}} |\nabla \Delta \varphi|^2 + \int_0^T \int_{\omega_p} e^{\frac{-M_0}{T-t}} |\Delta \psi|^2 \right.$$

$$\left. + \int_0^T \int_\Omega e^{\frac{-M_0}{T-t}} |-\beta_2 \nabla \psi_t - \nabla \Delta \psi + \beta_1 \nabla \Delta \varphi|^2 + \int_0^T \int_\Omega e^{\frac{-M_0}{T-t}} |\Gamma \varphi_{tt} + \Delta^2 \varphi|^2 \right).$$

$$(3.16)$$

Once we have stated the observability results for the cascade system we continue with the proof of Proposition 3.4. First we use (3.7) to prove a controllability result for system (3.10) with two controls, in the parabolic and hyperbolic equations respectively, and then we show how we can remove the control acting in the plate equation.

Following [4] we write the controllability problem in a variational form. To this end we introduce the following operators:

$$L(\varphi, \psi) = \begin{pmatrix} \Gamma\varphi_{tt} + \Delta^2\varphi + \beta_1\Delta\psi \\ \beta_2\psi_t - \Delta\psi \end{pmatrix}, \tag{3.17}$$

$$L^*(\varphi, \psi) = \begin{pmatrix} \Gamma\varphi_{tt} + \Delta^2\varphi \\ -\beta_2\psi_t - \Delta\psi + \beta_1\Delta\varphi \end{pmatrix}, \tag{3.18}$$

defined on the linear space

$$E_0 = \{(\varphi, \psi) \in C^4([0, T] \times \Omega) \times C^2([0, T] \times \Omega), \text{ such that}$$
$$\varphi, \psi = 0 \text{ in } \Sigma_T\}. \tag{3.19}$$

Note that the first two equations in system (3.10) can be written as follows,

$$L(\varphi, \psi) + \begin{pmatrix} 0 \\ G \end{pmatrix} = \begin{pmatrix} f_h \\ f_p \end{pmatrix}. \tag{3.20}$$

We now introduce the bilinear form

$$<(\varphi, \psi), (\hat{\varphi}, \hat{\psi})> = \int_0^T \int_\Omega e^{\frac{-M_0}{T-t}} L^*(\varphi, \psi) \cdot L^*(\hat{\varphi}, \hat{\psi})$$
$$+ \iint_Q e^{\frac{-M_0}{T-t}} \left(\nabla\Delta(\rho_h\varphi) \cdot \nabla\Delta(\rho_h\hat{\varphi}) + \Delta(\rho_p\psi)\Delta(\rho_p\hat{\psi}) \right), \tag{3.21}$$

for $(\varphi, \psi) \in E_0$. The weight functions ρ_h and ρ_p are cutoff functions that allows us to localize the action of the controls on ω_h and ω_p, respectively. More precisely, consider $\omega_{1,h} \subset \omega_h$, $\omega_{1,p} \subset \omega_p$ with $\text{dist}(\partial\omega_{1,h} \cap \Omega, \partial\omega_h \cap \Omega) \geq a > 0$, $\text{dist}(\partial\omega_{1,p}, \partial\omega_p) \geq a > 0$, for some $a > 0$, and $\omega_{1,h}$ a neighborhood of $M(x_0)$. We define

$$\rho_h, \rho_p \in C^2(\mathbb{R}^2); \quad 1 \geq \rho_h, \ \rho_p \geq 0 \tag{3.22}$$

such that

$$\begin{cases} \rho_h(x) = 0 & \text{for } x \in \Omega\backslash\omega_h \\ \rho_h(x) = 1 & \text{for } x \in \omega_{1,h} \end{cases} \qquad \begin{cases} \rho_p(x) = 0 & \text{for } x \in \Omega\backslash\omega_p \\ \rho_p(x) = 1 & \text{for } x \in \omega_{1,p} \end{cases}. \tag{3.23}$$

It is easy to see that the bilinear form defined in (3.21) is a scalar product in E_0. In fact, this is a consequence of the fact that L^* is in cascade form and the unique continuation for the plate and heat equations.

Let E the completion of E_0 with respect to this scalar product. Then E is a Hilbert space with this scalar product.

Now, given $v_0 \in H_D^0$, $v_1 \in H_D^{-1}$, $\eta_0 \in H_D^{-2}$, $G \in L^2(e_M; H_D^{-3})$ we define the linear form:

$$\mathcal{L}(\varphi, \psi) = -\beta_2 <\eta_0, \psi(0)>_{H_D^{-2}, H_D^2} - <v_1, \varphi(0)>_{H_D^{-1}, H_D^1}$$

$$+ \gamma <v_1, \Delta\varphi(0)>_{H_D^{-1}, H_D^1} + \int_\Omega v_0 \varphi_t(0) - \gamma \int_\Omega v_0 \Delta\varphi_t(0)$$

$$- \int_0^T <G, \psi>_{H_D^{-3}, H_D^3}. \tag{3.24}$$

From Proposition 3.7, $\mathcal{L}$ is continuous in E and therefore the problem

$$<(\varphi, \psi), (\hat\varphi, \hat\psi)> = \mathcal{L}(\hat\varphi, \hat\psi), \quad \text{for all } (\hat\varphi, \hat\psi) \in E, \tag{3.25}$$

admits a unique solution $(\varphi, \psi) \in E$, by Riesz's representation theorem. Therefore,

$$\begin{pmatrix} v \\ \eta \end{pmatrix} = e^{\frac{-M_0}{T-t}} L^*(\varphi, \psi), \tag{3.26}$$

is a controlled solution of (3.10) with controls

$$f_h = e^{\frac{-M_0}{T-t}} \Delta^3(\rho_h \varphi) \in C(0, T; H_D^{-1}), \tag{3.27}$$

$$f_p = e^{\frac{-M_0}{T-t}} \Delta^2(\rho_h \psi) \in C(0, T; H_D^{-2}). \tag{3.28}$$

Moreover, it is clear that the parabolic part of the controlled solution η satisfies (3.12) for any $M < M_0$. On the other hand, by considering (3.25) with $(\hat\varphi, \hat\psi) = (\varphi, \psi)$ we easily deduce the estimate,

$$\|f_p\|_{L^2(0,T;H_D^{-2})} + \|f_h\|_{L^2(0,T;H_D^{-3})} \le C(\|(v_0, v_1, \eta_0)\|_{V'} + \|G\|_{L^2(0,T;H_D^{-3})}). \tag{3.29}$$

It remains to prove that we can remove the control f_h. To this end we define $\omega = \omega_h$ and we consider the case $\omega_p = \omega$. In this way we have both controls supported in the same open subset ω. We refer to these controls as g_p and g_h.

Note that we can write

$$g_h = \Delta\tilde g_h, \quad \text{with } \tilde g_h = e^{\frac{-M_0}{T-t}} \Delta^2(\rho_h \varphi) \in C(0, T; H_D^{-1}), \tag{3.30}$$

and $\tilde{g}_h$ is supported in $\omega \times (0, T)$. Moreover,

$$(\tilde{g}_h)_t = e^{\frac{-M_0}{T-t}} \Delta^2(\rho_h \varphi_t) - \frac{M_0}{(T-t)^2} e^{\frac{-M_0}{T-t}} \Delta^2(\rho_h \varphi) \in C(0, T; H_D^{-2}). \quad (3.31)$$

Therefore, if we define

$$\tilde{\eta} = \eta - \frac{1}{\beta_1} \tilde{g}_h \in C(0, T; H_D^{-2}) \cap L^2(e_M, H_D^{-2}),$$

we can easily check that $(v, \tilde{\eta})$ is a controlled solution of (3.10) with $f_h = 0$ and

$$f_p = g_p - \frac{\beta_2}{\beta_1} \tilde{g}_h + \frac{1}{\beta_1} g_h,$$

which is supported in $\omega \times (0, T)$. Using estimate (3.29) for g_p and g_h we can easily check that f_p satisfies (3.12). A straightforward computation shows that $\tilde{\eta}$ satisfies (3.11) for any $M < M_0$.

3.4 Appendix

In this section we prove Lemma 3.5.

Assume that ω_h is a neighborhood of a set $M(x_0)$ for some x_0 in $\mathbb{R}$ and let $T > 2\sqrt{\gamma} R(x_0)$. Consider the following plate model,

$$\begin{cases} \Gamma \varphi_{tt} + \Delta^2 \varphi = F, & \text{in } \Omega, \\ \varphi = \Delta \varphi = 0, & \text{on } \partial\Omega, \\ \varphi(0) = \varphi^0, \varphi_t(0) = \varphi^1 & \text{in } \Omega, \end{cases} \quad (3.32)$$

where (φ^0, φ^1) are some initial data and F a second-hand term.

The inequality (3.13) can be easily deduced from the following two inequalities:

$$\mathcal{E}(0) \leq C \left(\int_0^T \int_{\omega_h} e^{\frac{-M_0}{T-t}} |\nabla \Delta \varphi|^2 + \int_0^T \int_\Omega e^{\frac{-M_0}{T-t}} |F|^2 \right), \quad (3.33)$$

$$\int_0^T \int_\Omega e^{\frac{-M_0}{T-t}} |\nabla \Delta \varphi(x, t)|^2 dx$$
$$\leq C \left(\int_0^T \int_{\omega_h} e^{\frac{-M_0}{T-t}} |\nabla \Delta \varphi|^2 + \int_0^T \int_\Omega e^{\frac{-M_0}{T-t}} |F|^2 \right), \quad (3.34)$$

for some constant $C > 0$. We prove separately both inequalities.

Step 1: Proof of inequality (3.33). When $F = 0$, there exists $C > 0$ such that

$$\mathcal{E}_\varphi(0) \leq C \left(\int_0^T \int_{\omega_h} |\nabla \Delta \varphi|^2 \right) \tag{3.35}$$

for any solution of (3.32) with initial data $(\varphi^0, \varphi^1) \in H_D^3 \times H_D^2$ (see [5]). As (3.35) holds for all $T > 2\sqrt{\gamma} R(x_0)$ we easily deduce

$$\mathcal{E}_\varphi(0) \leq C \left(\int_0^T \int_{\omega_h} e^{-\frac{M}{T-t}} |\nabla \Delta \varphi|^2 \right), \tag{3.36}$$

for another constant $C > 0$. We just write (3.35) for a smaller T^* such that $T > T^* > 2\sqrt{\gamma} R(x_0)$ and use the fact that $e^{-\frac{M}{T-t}} > \alpha > 0$ for some constant α and for all $t \in [0, T^*]$.

We have to prove that when $F \neq 0$ estimate (3.36) still holds but adding a suitable term to the right-hand side, involving F.

We decompose $\varphi = \phi^1 + \phi^2$ where ϕ^1 and ϕ^2 solve

$$\begin{cases} \Gamma \phi_{tt}^1 + \Delta^2 \phi^1 = F, & \text{in } \Omega, \\ \phi^1 = \Delta \phi^1 = 0, & \text{on } \partial\Omega, \\ \phi^1(0) = 0, \phi_t^1(0) = 0, & \text{in } \Omega, \end{cases} \tag{3.37}$$

$$\begin{cases} \Gamma \phi_{tt}^2 + \Delta^2 \phi^2 = 0, & \text{in } \Omega, \\ \phi^2 = \Delta \phi^2 = 0, & \text{on } \partial\Omega, \\ \phi^2(0) = \varphi^0, \phi_t^2(0) = \varphi^2, & \text{in } \Omega. \end{cases} \tag{3.38}$$

From Duhamel's principle, we can write

$$\phi^1(\cdot, t) = \int_0^t \psi(\cdot, t - s, s) \, ds, \tag{3.39}$$

where $\psi(\cdot, t - s, s)$ solves, for each value of the parameter $s \in (0, t)$,

$$\begin{cases} \Gamma \psi_{tt}^2 + \Delta^2 \psi = 0, & \text{in } \Omega, \\ \psi = \Delta \psi^2 = 0, & \text{on } \partial\Omega, \\ \psi(\cdot, 0, s) = 0, \psi_t(\cdot, 0, s) = F(\cdot, s) & \text{in } \Omega. \end{cases} \tag{3.40}$$

Using (3.39) and the fact that

$$e^{-\frac{M}{T-t}} \leq e^{-\frac{M}{T-s}}, \quad \text{when } 0 < s \leq t < T,$$

we have

$$e^{-\frac{M}{T-t}} |\phi^1(\cdot, t)| \leq \int_0^t e^{-\frac{M}{T-s}} |\psi(\cdot, t - s, s)| \, ds.$$

Therefore, by the continuous dependence with respect to the initial datum for solutions of (3.40), we have

$$\|e^{-\frac{M}{T-t}}\phi^1\|^2_{L^2(0,T;H^3_D)} \leq \int_0^T \|e^{-\frac{M}{T-s}}\psi(\cdot,\cdot,s)\|^2_{L^2(0,T;H^3_D)} ds$$

$$\leq C\int_0^T e^{-\frac{M}{T-s}}\|(\psi(\cdot,0,s),\psi_t(\cdot,0,s))\|^2_{H^3_D \times H^2_D} ds$$

$$= \|e^{-\frac{M}{T-s}}F(x,s)\|^2_{L^2(0,T;H^2_D)}. \tag{3.41}$$

Combining (3.41) and the estimate (3.36) for ϕ^2 we easily obtain

$$\mathcal{E}_\varphi(0) = \mathcal{E}_{\phi^2}(0) \leq C\left(\int_0^T\!\!\int_{\omega_h} e^{-\frac{M}{T-t}}|\nabla\Delta\phi^2|^2\right)$$

$$\leq C\left(\int_0^T\!\!\int_{\omega_h} e^{-\frac{M}{T-t}}|\nabla\Delta\varphi|^2 + e^{-\frac{M}{T-t}}|\nabla\Delta\phi^1|^2\right)$$

$$\leq C\left(\int_0^T\!\!\int_{\omega_h} e^{-\frac{M}{T-t}}|\nabla\Delta\varphi|^2 + \|e^{-\frac{M}{T-t}}F\|^2_{L^2(0,T;H^2_D)}\right). \tag{3.42}$$

This completes the proof of inequality (3.33).

Step 2: Proof of inequality (3.34). As in the previous step, we decompose $\varphi = \phi^1 + \phi^2$ where ϕ^1 and ϕ^2 solve (3.37) and (3.38), respectively. Thus, it suffices to prove inequality (3.34) when we consider both ϕ^1 and ϕ^2 on the left-hand side.

The inequality (3.34) for ϕ^1 is in fact (3.41).

Concerning the inequality for ϕ^2, it is easily reduced to the one proved in Step 1. In fact, the solutions of (3.38) conserve the energy, i.e.,

$$\mathcal{E}_{\phi^2}(t) = \mathcal{E}_{\phi^2}(0), \quad t \in [0,T],$$

and therefore

$$\int_0^T\!\!\int_\Omega e^{\frac{-M_0}{T-t}}|\nabla\Delta\phi^2(x,t)|^2 dx \leq \int_0^T e^{\frac{-M_0}{T-t}}\mathcal{E}_{\phi^2}(t)dt$$

$$= \mathcal{E}_{\phi^2}(0)\int_0^T e^{\frac{-M_0}{T-t}} dt = C\mathcal{E}_\varphi(0).$$

This inequality together with the one in Step 1 provides (3.34) when we consider ϕ^2, instead of φ, on the left-hand side.

References

[1] G. Avalos, Exact controllability of a thermoelastic system with control in the thermal component only, *Diff. Integr. Eq.* 13 (4–6) (2000), 613–30.

[2] G. Avalos, I. Lasiecka, The null controllability of thermoelastic plates and singularity of the associated minimal energy function, *Math. Anal. Appl.* 294 (2004), 34–61.

[3] A. Benabdallah, M.G. Naso, Null controllability of a thermoelastic plate, *Abstr. Appl. Anal.* 7 (11) (2002), 585–99.

[4] E. Fernández-Caray, A. Munch, Strong convergent approximations of null controls for the heat equation, *SEMA J.* 61 (1) (2013), 49–78.

[5] C. Castro, L. de Teresa, Null controllability of the linear system of thermoelastic plates, *J. Math. Anal. App.* 428 (2) (2015), 772–93.

[6] M. Eller, I. Lasiecka, R. Triggiani, Simultaneous exact/approximate boundary controllability of thermo-elastic plates with variable thermal coefficient and moment control, *J. Math. Anal. App.* 251 (2000), 452–78.

[7] A. Fursikov, O.Yu. Imanuvilov, *Controllability of Evolution Equations*, Lecture Notes Ser. 34, Seoul National University, Korea, 1996.

[8] S.W. Hansen, B. Zhang, Boundary control of a linear thermo-elastic beam, *J. Math. Anal. Appl.*, 210 (1997), 182–205.

[9] E. Lagnese, The reachability problem of thermoelastic plates, *Arch. Rat. Mech. Anal.* 112 (1990), 223–67.

[10] I. Lasiecka, T.I. Seidman, Blowup estimates for observability of a thermoelastic system, *Asymptot. Anal.* 50 (1–2) (2006), 93–120.

[11] I. Lasiecka, R. Triggiani, Exact null controllability of structurally damped and thermoelastic parabolic models, *Rend. Mat. Acc. Lincei* 9 (9) (1998), 43–69.

[12] L. de Teresa, E. Zuazua, Controllability of the linear system of thermoelastic plates, *Adv. Diff. Eq.* 1 (3) (1996), 369–402.

DEPARTAMENTO DE MATEMÁTICA E INFORMÁTICA, ETSI CAMINOS, CANALES Y PUERTOS, UNIVERSIDAD POLITÉCNICA DE MADRID, 28040 MADRID, SPAIN

E-mail address: carlos.castro@upm.es

INSTITUTO DE MATEMÁTICAS, UNIVERSIDAD NACIONAL AUTÓNOMA DE MÉXICO, MÉXICO

E-mail address: deteresa@matem.unam.mx

4

Doubly Connected V-States for the Generalized Surface Quasi-geostrophic Equations

FRANCISCO DE LA HOZ, ZINEB HASSAINIA,
AND TAOUFIK HMIDI

Abstract

In this chapter, we prove the existence of doubly connected V-states for the generalized SQG equations with $\alpha \in \,]0, 1[$. They can be described by countable branches bifurcating from the annulus at some explicit "eigenvalues" related to Bessel functions of the first kind. Contrary to Euler equations [Hmidi et al., *SIAM J. Math. Anal.* 48 (2016), no. 3, 1892–928], we find V-states rotating with positive and negative angular velocities. At the end of the paper we discuss some numerical experiments concerning the limiting V-states.

1991 Mathematics Subject Classification. 35Q35, 76B03, 76C05
Key words and phrases. gSQG equations, V-states, doubly connected patches, bifurcation theory

Contents

4.1 Introduction and Main Result

This note is a short version of the paper [22] which deals with the generalized surface quasi-geostrophic (gSQG) equation arising in fluid dynamics. This model describes the evolution of the potential temperature θ by the nonlinear transport equation:

$$\begin{cases} \partial_t\theta + u \cdot \nabla\theta = 0, & (t,x) \in \mathbb{R}_+ \times \mathbb{R}^2, \\ u = -\nabla^\perp(-\Delta)^{-1+\frac{\alpha}{2}}\theta, \\ \theta_{|t=0} = \theta_0. \end{cases} \tag{4.1}$$

Here u refers to the velocity field, $\nabla^\perp = (-\partial_2, \partial_1)$, and α is a real parameter taken in $]0, 2[$. The singular operator $(-\Delta)^{-1+\frac{\alpha}{2}}$ is of convolution type and defined by,

$$(-\Delta)^{-1+\frac{\alpha}{2}}\theta(x) = \frac{C_\alpha}{2\pi} \int_{\mathbb{R}^2} \frac{\theta(y)}{|x-y|^\alpha} dy, \tag{4.2}$$

with $C_\alpha = \frac{\Gamma(\alpha/2)}{2^{1-\alpha}\Gamma(\frac{2-\alpha}{2})}$ where Γ stands for the gamma function. This model was proposed by Córdoba et al. [11] as an interpolation between Euler equations and the surface quasi-geostrophic (SQG) model corresponding to $\alpha = 0$ and $\alpha = 1$, respectively. The SQG equation was used by Juckes [23] and Held et al. [17] to describe the atmosphere circulation near the tropopause. It was also used by Lapeyre and Klein [28] to track the ocean dynamics in the upper layers. We note that there is a strong mathematical and physical analogy with the three-dimensional incompressible Euler equations (see [10] for details).

In the last few years there has been a growing interest in the mathematical study of these active scalar equations. Special attention has been paid to the local well-posedness of classical solutions which can be performed in various functional spaces. For instance, this was implemented in the framework of Sobolev spaces [7] by using the commutator theory. Whether or not these solutions are global in time is an open problem except for Euler equations $\alpha = 0$. The second restriction with the gSQG equation concerns the construction of Yudovich solutions – known to exist globally in time for Euler equations [34] – which are not at all clear even locally in time. The main difficulty is due to the velocity, which is in general singular and scales below the Lipschitz class. Nonetheless, one can say more about this issue for some special class of concentrated vortices. More precisely, when the initial datum has a vortex patch structure, that is, $\theta_0(x) = \chi_D$ is the characteristic function of a bounded simply connected smooth domain D, then there is a unique local solution in the patch form $\theta(t) = \chi_{D_t}$. In this case, the boundary motion of the domain D_t is described by the contour dynamics formulation; see the papers [15, 32]. The global persistence of the boundary regularity is only known for $\alpha = 0$ according to Chemin's [9] result; for another proof see the paper of Bertozzi and Constantin [2]. Notice that for $\alpha > 0$ the numerical experiments carried out in [11] provide strong evidence for the singularity formation in finite time. Let us mention that the contour dynamics equation remains locally well-posed when the domain of the initial patch is assumed to be multiconnected, meaning that the boundary is composed with finite number of disjoint smooth Jordan curves.

The main concern of this work is to explore analytically and numerically some special vortex patches called V-states; they correspond to patches which do not change their shapes during the motion. The emphasis will be put on the V-states subject to uniform rotation around their center of mass, that is, $D_t = \mathbf{R}_{x_0, \Omega t} D$, where $\mathbf{R}_{x_0, \Omega t}$ stands for the planar rotation with center x_0 and angle Ωt. The parameter Ω is called the angular velocity of the rotating domain. Along this note we call these structures rotating patches or simply V-states. Their existence is of great interest for at least two reasons: first, they provide nontrivial initial data with global existence and second, this might explain the emergence of some ordered structures in the geophysical flows. This study has been conducted first for the two-dimensional Euler equations ($\alpha = 0$) a long time ago and a number of analytical and numerical studies are known in the literature. The first result in this setting goes back to Kirchhoff [26] who discovered that an ellipse of semi-axes a and b rotates uniformly with the angular velocity $\Omega = ab/(a+b)^2$ (see for instance the references [3, p. 304] and [27, p. 232]). Till now this is the only known explicit V-states; however the existence of implicit examples was established about one century later. In fact, Deem and Zabusky [13] gave numerical evidence of the existence of the V-states with m-fold symmetry for each integer $m \geq 2$; note that the case $m = 2$ coincides with Kirchhoff's ellipses. To fix the terminology, a planar domain is said to be m-fold symmetric if it has the same group invariance of a regular polygon with m sides. Note that at each frequency m these V-states can be seen as a continuous deformation of the disk with respect to to the angular velocity. An analytical proof of this fact was given a few years later by Burbea [4]. His approach consists in writing a stationary problem in the frame of the patch with the conformal mapping of the domain and to look for the nontrivial solutions by using the technique of the bifurcation theory. Quite recently, in [20] Burbea's approach was revisited with more details and explanations. The boundary regularity of the V-states was also studied and it was shown to be of class C^∞ and convex close to the disk.

We mention that explicit vortex solutions similar to the ellipses are discovered in the literature for the incompressible Euler equations in the presence of an external shear flow; see for instance [8, 24, 29]. A general review about vortex dynamics can be found in the papers [1, 30].

With regard to the existence of the simply connected V-states for the (gSQG) it has been discussed very recently in the papers [5, 6, 16]. In [5], it was shown that the ellipses cannot rotate for any $\alpha \in (0,2)$ and to the authors' best knowledge no explicit example is known in the literature. Lately, in [16] the last two authors proved the analogous of Burbea's result and showed the existence of the m-folds rotating patches for $\alpha \in]0,1[$. In addition, the bifurcation from the unit disk occurs at the angular velocities,

$$\Omega_m^\alpha \triangleq \frac{\Gamma(1-\alpha)}{2^{1-\alpha}\Gamma^2(1-\frac{\alpha}{2})}\left(\frac{\Gamma(1+\frac{\alpha}{2})}{\Gamma(2-\frac{\alpha}{2})} - \frac{\Gamma(m+\frac{\alpha}{2})}{\Gamma(m+1-\frac{\alpha}{2})}\right), \quad m \geq 2,$$

where Γ denotes the usual gamma function. The remaining case $\alpha \in [1,2)$ has been explored and solved by Castro et al. [6]. They also show that the V-states are C^∞ and convex close to the disks.

We want in this paper to learn more about the V-states but with different topological structure compared to the preceding discussion. More precisely, we propose to scrutinize rotating patches with only one hole, also called doubly connected V-states. Recall that a patch $\theta_0 = \chi_D$ is said to be doubly connected if the domain $D = D_1 \backslash D_2$, with D_1 and D_2 being two simply connected bounded domains satisfying $\overline{D_2} \subset D_1$. This structure is preserved for Euler system globally in time but known to be for short time when $\alpha \in]0,1[$ (see [7, 15, 32]). We notice that compared to the simply connected case the boundaries evolve through extra nonlinear terms coming from the interaction between the boundaries and therefore the existence of the V-states is relatively more complicated to analyze. This problem is not well studied from the analytical point of view and recent progress has been made for Euler equations in the papers [14, 19, 21]. In [21], the authors proved that the existence of explicit V-states similar to Kirchhoff ellipses seems to be out of reach. Indeed, it was stated that if one of the boundaries of the V-state is a circle then necessarily the other one should be also a circle. Moreover, if the inner curve is an ellipse then there is no rotation at all. Another closely related subject is to deal with some vortex magnitude μ inside the domain D_2 and try to find explicit rotating patches. This was done by Flierl and Polvani [14] who proved that confocal ellipses rotate uniformly provided some compatibility relations are satisfied between the parameter μ and the semi-axes of the ellipses. We note that another approach based upon complex analysis tools with a complete discussion can be found in [21].

Now, from (4.1) we may easily conclude that the annulus is a stationary doubly connected patch, and therefore it rotates with any angular velocity Ω. From this obvious fact, one can wonder wether or not the bifurcation to nontrivial V-states still happens as for the simply connected case. This has been recently investigated in [19] for Euler equations following basically Burbea's approach but with more involved calculations. It was shown that for $b \in (0, 1)$ and m being an integer satisfying the inequality

$$m > \frac{2(1 + b^m)}{1 - b^2} \qquad (4.3)$$

then there exist two curves of nonannular m-fold doubly connected patches bifurcating from the annulus $\{z; b < |z| < 1\}$ at different eigenvalues $\Omega_m^{\pm}$ given explicitly by the formula

$$\Omega_m^{\pm} = \frac{1 - b^2}{4} \pm \frac{1}{2m} \sqrt{\left[\frac{m}{2}(1 - b^2) - 1\right]^2 - b^{2m}}.$$

Now we come to the main contribution of the current work. We propose to study the doubly connected V-states for the gSQG model (4.1) when $\alpha \in \,]0, 1[$. Before stating our result we need to make some notation. We define

$$\Lambda_n(b) \triangleq \frac{1}{b} \int_0^{+\infty} J_n(bt) J_n(t) \frac{dt}{t^{1-\alpha}},$$

and

$$\Theta_n \triangleq \Lambda_1(1) - \Lambda_n(1),$$

where J_n refers to Bessel function of the first kind. Our result reads as follows.

Theorem 4.1 *Let $\alpha \in [0, 1[$ and $b \in \,]0, 1[$; there exists $N \in \mathbb{N}$ with the following property.*

For each $m > N$ there exists two curves of m-fold doubly connected V-states that bifurcate from the annulus $\{z \in \mathbb{C}, b < |z| < 1\}$ at the angular velocities

$$\Omega_m^{\alpha,\pm} \triangleq \frac{1 - b^2}{2} \Lambda_1(b) + \frac{1}{2}(1 - b^{-\alpha})\Theta_m \pm \frac{1}{2}\sqrt{\Delta_m(\alpha, b)},$$

with

$$\Delta_m(\alpha, b) \triangleq \left[(b^{-\alpha} + 1)\Theta_m - (1 + b^2)\Lambda_1(b) \right]^2 - 4b^2 \Lambda_m^2(b).$$

Remarks

1. The number N is the smallest integer such that

$$\Theta_N \geq \frac{1 + b^2}{b^{-\alpha} + 1}\Lambda_1(b) + \frac{2b}{b^{-\alpha} + 1}\Lambda_N(b). \qquad (4.4)$$

2. We can check by using the strict monotonicity of $b \mapsto \Lambda_1(b)$ that for any $b, \alpha \in (0, 1)$,

$$\lim_{m \to +\infty} \Omega_m^{\alpha, -} = -b^{-\alpha}\Lambda_1(1) + \Lambda_1(b) < 0.$$

Consequently the corresponding bifurcating curves generate close to the annulus nontrivial clockwise doubly connected V-states. This fact is completely new compared to what we know for Euler equations or for the simply connected case where the bifurcation occurs at positive angular velocities. The numerical experiments reveal the existence of non radial stationary patches for the gSQG equations and it would be very interesting to establish this fact analytically. In a connected subject, we point out that the last author has shown quite recently in [18] that for Euler equations clockwise convex V-states reduce to the disks.

4.2 Boundary Equations

In what follows we shall write down the equation governing the boundary of the doubly connected V-states. Let $D = D_1 \backslash D_2$ be a doubly connected domain, that is, D_1 and D_2 are two simply connected domains with $D_2 \subset D_1$. Denote by Γ_1 and Γ_2 their boundaries, respectively. Consider the parametrization by the conformal mapping: $\phi_j : \mathbb{D}^c \to D_j^c$ satisfying

$$\phi_1(z) = z + f_1(z) = z\left(1 + \sum_{n=1}^{\infty} \frac{a_n}{z^n}\right),$$

and

$$\phi_2(z) = bz + f_2(z) = z\left(b + \sum_{n=1}^{\infty} \frac{b_n}{z^n}\right), \quad 0 < b < 1.$$

Now assume that $\theta_0 = \chi_D$ is a rotating patch for the model (4.1) then according to [16] the boundary equations are given by

$$\Omega \operatorname{Re}\{z\overline{z'}\} = C_\alpha \operatorname{Im}\left\{\frac{1}{2\pi}\int_{\partial D}\frac{d\zeta}{|z-\zeta|^\alpha}\overline{z'}\right\}, \qquad \forall z \in \partial D = \Gamma_1 \cup \Gamma_2.$$

$$= C_\alpha \operatorname{Im}\left\{\left(\frac{1}{2\pi}\int_{\Gamma_1}\frac{d\zeta}{|z-\zeta|^\alpha} - \frac{1}{2\pi}\int_{\Gamma_2}\frac{d\zeta}{|z-\zeta|^\alpha}\right)\overline{z'}\right\}, \quad (4.5)$$

where z' denotes a tangent vector to the boundary ∂D at the point z. We shall now rewrite the equations by using the conformal parametrizations ϕ_1 and ϕ_2. First remark that for $w \in \mathbb{T}$ a tangent vector on the boundary Γ_j at the point $\phi_j(w)$ is given by

$$\overline{z'} = -i\overline{w}\,\overline{\phi'_j(w)}.$$

Inserting this into (4.5) and using the change of variables $\tau = \phi_j(w)$ give

$$\forall w \in \mathbb{T}, \quad F_j(\Omega, \phi_1, \phi_2)(w) = 0; \quad j=1,2,$$

with

$$F_j(\Omega, \phi_1, \phi_2)(w) \triangleq \Omega \operatorname{Im}\left\{\phi_j(w)\overline{w}\,\overline{\phi'_j(w)}\right\}$$
$$+ C_\alpha \operatorname{Im}\left\{\left(\fint_{\mathbb{T}}\frac{\phi'_2(\tau)d\tau}{|\phi_j(w)-\phi_2(\tau)|^\alpha} - \fint_{\mathbb{T}}\frac{\phi'_1(\tau)d\tau}{|\phi_j(w)-\phi_1(\tau)|^\alpha}\right)\overline{w}\,\overline{\phi'_j(w)}\right\}$$
$$(4.6)$$

and $C_\alpha \triangleq \dfrac{\Gamma(\alpha/2)}{2^{1-\alpha}\Gamma(\frac{2-\alpha}{2})}$. We shall introduce the functionals

$$G_j(\Omega, f_1, f_2) \triangleq F_j(\Omega, \phi_1, \phi_2) \quad j=1,2.$$

Then equations of the V-states become,

$$\forall w \in \mathbb{T}, \quad G_j(\Omega, f_1, f_2)(w) = 0, \quad j=1,2. \quad (4.7)$$

Now it is easy to ascertain from straightforward calculus that the annulus is a rotating patch for any $\Omega \in \mathbb{R}$. This means that

$$\forall w \in \mathbb{T}, \quad F_j(\Omega, \operatorname{Id}, b\operatorname{Id})(w) = 0, \quad \text{for } j \in \{1,2\}.$$

4.3 Basic Tools

In this section we shall recall in the first part some simple facts about Hölder spaces on the unit circle $\mathbb{T}$. In the second part we state

Crandall–Rabinowitz's theorem which is a crucial tool in the proof of Theorem 4.1. We end with some important properties of the hypergeometric functions which appear in a natural way in the spectral study of the linearized operator.

4.3.1 Functional Spaces

We shall introduce Hölder spaces on the unit circle $\mathbb{T}$. Let $0 < \gamma < 1$. We denote by $C^\gamma(\mathbb{T})$ the space of continuous functions f such that

$$\|f\|_{C^\gamma(\mathbb{T})} \triangleq \|f\|_{L^\infty(\mathbb{T})} + \sup_{\tau \neq w \in \mathbb{T}} \frac{|f(\tau) - f(w)|}{|\tau - w|^\alpha} < \infty.$$

For any integer n, the space $C^{n+\gamma}(\mathbb{T})$ stands for the set of functions f of class C^n whose n-th order derivatives are Hölder continuous with exponent γ. It is equipped with the usual norm,

$$\|f\|_{C^{n+\gamma}(\mathbb{T})} \triangleq \|f\|_{L^\infty(\mathbb{T})} + \left\| \frac{d^n f}{dw^n} \right\|_{C^\gamma(\mathbb{T})}.$$

Recall that the Lipschitz seminorm is defined by,

$$\|f\|_{\mathrm{Lip}(\mathbb{T})} = \sup_{\tau \neq w \in \mathbb{T}} \frac{|f(\tau) - f(w)|}{|\tau - w|}.$$

Now we list some classical properties that will be used later at many places.

1. For $n \in \mathbb{N}, \gamma \in]0, 1[$ the space $C^{n+\gamma}(\mathbb{T})$ is an algebra.
2. For $K \in L^1(\mathbb{T})$ and $f \in C^{n+\gamma}(\mathbb{T})$ we have the convolution law,

$$\|K * f\|_{C^{n+\gamma}(\mathbb{T})} \leq \|K\|_{L^1(\mathbb{T})} \|f\|_{C^{n+\gamma}(\mathbb{T})}.$$

4.3.2 Crandall–Rabinowitz's Theorem

We shall recall Crandall–Rabinowitz's theorem which is a basic tool of the bifurcation theory and will be used in the proof of Theorem 4.1. For the proof, we refer for instance to [12, 25].

Theorem 4.2 *Let X, Y be two Banach spaces, V a neighborhood of 0 in X and let $F: \mathbb{R} \times V \to Y$ with the following properties:*

1. *$F(\lambda, 0) = 0$ for any $\lambda \in \mathbb{R}$.*
2. *The partial derivatives F_λ, F_x and $F_{\lambda x}$ exist and are continuous.*
3. *The kernel $N(\mathcal{L}_0)$ is one-dimensional and the range $R(\mathcal{L}_0)$ is of codimension one.*

4. *Transversality assumption:* $\partial_\lambda \partial_x F(0,0)x_0 \notin R(\mathcal{L}_0)$, *where*

$$N(\mathcal{L}_0) = span\{x_0\}, \quad \mathcal{L}_0 \triangleq \partial_x F(0,0).$$

If Z is any complement of $N(\mathcal{L}_0)$ *in X, then there is a neighborhood U of* $(0,0)$ *in* $\mathbb{R} \times X$, *an interval* $(-a,a)$, *and continuous functions* $\varphi : (-a,a) \to \mathbb{R}$, $\psi : (-a,a) \to Z$ *such that* $\varphi(0) = 0$, $\psi(0) = 0$ *and*

$$F^{-1}(0) \cap U = \left\{ (\varphi(\xi), \xi x_0 + \xi \psi(\xi)); \ |\xi| < a \right\} \cup \left\{ (\lambda, 0); \ (\lambda, 0) \in U \right\}.$$

4.3.3 Special Functions

We shall give a short introduction on the Gauss hypergeometric functions and discuss some of their basic properties. The formulae listed below will be crucial in the computations of the linearized operator associated with the V-states equations. Recall that for any real numbers $a, b \in \mathbb{R}$, $c \in \mathbb{R} \backslash (-\mathbb{N})$ the hypergeometric function $z \mapsto F(a, b; c; z)$ is defined on the open unit disk $\mathbb{D}$ by the power series

$$F(a, b; c; z) = \sum_{n=0}^{\infty} \frac{(a)_n (b)_n}{(c)_n} \frac{z^n}{n!}, \quad \forall z \in \mathbb{D}.$$

Here, $(x)_n$ is the Pochhammer symbol defined by,

$$(x)_n = \begin{cases} 1 & n = 0 \\ x(x+1) \cdots (x+n-1) & n \geq 1. \end{cases}$$

It is obvious that

$$(x)_n = x(1+x)_{n-1}, \quad (x)_{n+1} = (x+n)(x)_n. \tag{4.8}$$

For future use we recall an integral representation of the hypergeometric function (for instance see [31, p. 47]). Assume that $\text{Re}(c) > \text{Re}(b) > 0$, then

$$F(a, b; c; z) = \frac{\Gamma(c)}{\Gamma(b)\Gamma(c-b)} \int_0^1 x^{b-1}(1-x)^{c-b-1}(1-zx)^{-a}dx, \quad |z| < 1. \tag{4.9}$$

The function $\Gamma : \mathbb{C} \backslash \{-\mathbb{N}\} \to \mathbb{C}$ refers to the gamma function which is the analytic continuation to the negative half plane of the usual gamma function defined on the positive half-plane $\{\text{Re } z > 0\}$ by the integral representation:

$$\Gamma(z) = \int_0^{+\infty} t^{z-1} e^{-t} dt.$$

It satisfies the relation

$$\Gamma(z+1) = z\,\Gamma(z), \quad \forall z \in \mathbb{C}\backslash(-\mathbb{N}). \tag{4.10}$$

From this we deduce the identities

$$(x)_n = \frac{\Gamma(x+n)}{\Gamma(x)}, \quad (x)_n = (-1)^n \frac{\Gamma(1-x)}{\Gamma(1-x-n)}, \tag{4.11}$$

provided all the quantities in the right terms are well-defined. Later we need the following values,

$$\Gamma(n+1) = n!, \quad \Gamma(1/2) = \sqrt{\pi}. \tag{4.12}$$

Another useful identity is Euler's reflection formula,

$$\Gamma(1-z)\Gamma(z) = \frac{\pi}{\sin(\pi z)}, \quad \forall z \notin \mathbb{Z}. \tag{4.13}$$

Now we shall introduce the digamma function which is nothing but the logarithmic derivative of the gamma function and often denoted by F. It is given by

$$F(z) = \frac{\Gamma'(z)}{\Gamma(z)}, \quad z \in \mathbb{C}\backslash(-\mathbb{N}).$$

The following identity is classical,

$$\forall n \in \mathbb{N}, \quad F\left(n+\frac{1}{2}\right) = -\gamma - 2\ln 2 + 2\sum_{k=0}^{n-1} \frac{1}{2k+1}. \tag{4.14}$$

When $\mathrm{Re}(c-a-b) > 0$ then it can be shown that the hypergeometric series is absolutely convergent on the closed unit disk and one has the expression,

$$F(a,b;c;1) = \frac{\Gamma(c)\Gamma(c-a-b)}{\Gamma(c-a)\Gamma(c-b)}. \tag{4.15}$$

The proof can be found in [31, p. 49].

Now recall the Kummer's quadratic transformation

$$F\left(a,b;2b;\frac{4z}{(1+z)^2}\right) = (1+z)^{2a} F\left(a, a+\frac{1}{2}-b; b+\frac{1}{2}; z^2\right), \quad \forall z \in [0,1[. \tag{4.16}$$

Next we recall some contiguous function relations of the hypergeometric series (see [31]).

$$c(c+1)F(a,b;c;z) - c(c+1)F(a,b;c+1;z)$$
$$-ab\,zF(a+1,b+1;c+2;z) = 0, \tag{4.17}$$

$$c\,F(a,b;c;z) - cF(a+1,b;c;z) + b\,zF(a+1,b+1;c+1;z) = 0,$$
$$(4.18)$$

$$c\,F(a,b;c;z) - cF(a,b+1;c;z) + a\,zF(a+1,b+1;c+1;z) = 0,$$
$$(4.19)$$

$$c\,F(a,b;c;z) - (c-b)F(a,b;c+1;z) - b\,F(a,b+1;c+1;z) = 0,$$
$$(4.20)$$

$$c\,F(a,b;c;z) - (c-a)F(a,b;c+1;z) - a\,F(a+1,b;c+1;z) = 0,$$
$$(4.21)$$

$$b\,F(a,b+1;c;z) - a\,F(a+1,b;c;z) + (a-b)F(a,b;c;z) = 0, \quad (4.22)$$

$$(b-a)(1-z)F(a,b;c;z) - (c-a)F(a-1,b;c;z)$$
$$+(c-b)F(a,b-1;c;z) = 0. \quad (4.23)$$

Now we recall the Bessel function J_n of the first kind of index $n \in \mathbb{N}$ and review some important identities. It is defined in the full space $\mathbb{C}$ by the power series

$$J_n(z) = \sum_{k \geq 0} \frac{(-1)^k}{k!\,(n+k)!} \left(\frac{z}{2}\right)^{2k+n}.$$

The following identity called Sonine–Schafheitlin's formula will be very useful later.

$$\int_0^{+\infty} \frac{J_\mu(at)J_\nu(bt)}{t^\lambda}\,dt = \frac{a^{\lambda-\nu-1}b^\nu\,\Gamma\left(\frac{1}{2}\mu + \frac{1}{2}\nu - \frac{1}{2}\lambda + \frac{1}{2}\right)}{2^\lambda\Gamma(\nu+1)\Gamma\left(\frac{1}{2}\lambda + \frac{1}{2}\mu - \frac{1}{2}\nu + \frac{1}{2}\right)}$$
$$\times F\left(\frac{\mu+\nu-\lambda+1}{2}, \frac{\nu-\lambda-\mu+1}{2}; \nu+1; \frac{b^2}{a^2}\right),$$
$$(4.24)$$

provided that $0 < b < a$ and that the integral is convergent. A detailed proof of this result can be found in [33, p. 401]. For a future use we introduce the quantity,

$$\Lambda_n(b) = \frac{\Gamma(\frac{\alpha}{2})}{2^{1-\alpha}\Gamma(1-\frac{\alpha}{2})} \frac{(\frac{\alpha}{2})_n}{n!} b^{n-1} F\left(\frac{\alpha}{2}, n+\frac{\alpha}{2}, n+1, b^2\right), \quad (4.25)$$

which is obviously positive for all $\alpha, b \in (0,1)$. From the relation (4.11) we get

$$\Lambda_n(b) = \frac{b^{n-1}}{2^{1-\alpha}\Gamma^2(1-\frac{\alpha}{2})} \int_0^1 x^{n-1+\frac{\alpha}{2}}(1-x)^{-\frac{\alpha}{2}}(1-b^2 x)^{-\frac{\alpha}{2}} \, dx, \quad b \in (0,1).$$
(4.26)

Therefore, it is easily seen that $b \in (0,1) \mapsto \Lambda_n(b)$ is increasing and $n \in \mathbb{N}^* \mapsto \Lambda_n(b)$ is decreasing. This implies in turn that $n \in \mathbb{N}^* \mapsto \Theta_n = \Lambda_1(1) - \Lambda_n(1)$ is increasing and thus it should be positive. Notice that these properties can be also proven from the series expansion (4.25).

We end this section with some integrals that will appear later in the spectral study. The proofs can be found in [22].

Lemma 4.3 Let $\alpha, b \in (0,1)$ and $n \in \mathbb{N}$. Then for any $w \in \mathbb{T}$ we have the following formulae:

$$\oint_{\mathbb{T}} \frac{\tau^{n-1}}{|w-b\tau|^{\alpha}} d\tau = w^n b^n \frac{(\frac{\alpha}{2})_n}{n!} F\left(\frac{\alpha}{2}, n+\frac{\alpha}{2}; n+1; b^2\right).$$
(4.27)

$$\oint_{\mathbb{T}} \frac{(w-b\tau)(aw^n - c\tau^n)}{|w-b\tau|^{\alpha+2}} d\tau = w^{n+2}b\left[a\left(1+\frac{\alpha}{2}\right)F\left(\frac{\alpha}{2}, 2+\frac{\alpha}{2}; 2; b^2\right)\right.$$
$$\left. - cb^n \frac{(1+\frac{\alpha}{2})_{n+1}}{(n+1)!} F\left(\frac{\alpha}{2}, n+2+\frac{\alpha}{2}; n+2; b^2\right)\right].$$
(4.28)

$$\oint_{\mathbb{T}} \frac{(\overline{w}-b\overline{\tau})(a\overline{w}^n - c\overline{\tau}^n)}{|w-b\tau|^{\alpha+2}} d\tau = \overline{w}^n\left[ab\frac{\alpha}{2}F\left(\frac{\alpha}{2}+1, \frac{\alpha}{2}+1; 2; b^2\right)\right.$$
$$\left. - cb^{n-1}\frac{(1+\frac{\alpha}{2})_{n-1}}{(n-1)!} F\left(\frac{\alpha}{2}, n+\frac{\alpha}{2}; n; b^2\right)\right].$$
(4.29)

$$\oint_{\mathbb{T}} \frac{(bw-\tau)(aw^n - c\tau^n)}{|bw-\tau|^{\alpha+2}} d\tau = -w^{n+2}b^2\left[a\frac{\alpha}{4}\left(\frac{\alpha}{2}+1\right)F\left(1+\frac{\alpha}{2}, 2+\frac{\alpha}{2}; 3; b^2\right)\right.$$
$$\left. - cb^n \frac{(\frac{\alpha}{2})_{n+2}}{(n+2)!} F\left(1+\frac{\alpha}{2}, n+2+\frac{\alpha}{2}; n+3; b^2\right)\right].$$
(4.30)

$$\oint_{\mathbb{T}} \frac{(b\overline{w}-\overline{\tau})(a\overline{w}^n - c\overline{\tau}^n)}{|bw-\tau|^{\alpha+2}} d\tau = -\overline{w}^n\left[aF\left(\frac{\alpha}{2}, \frac{\alpha}{2}+1; 1; b^2\right)\right.$$
$$\left. - cb^n \frac{(\frac{\alpha}{2})_n}{n!} F\left(\frac{\alpha}{2}+1, n+\frac{\alpha}{2}; n+1; b^2\right)\right].$$
(4.31)

4.4 Regularity of the Nonlinear Functional

This section is devoted to the regularity of the nonlinear functional G introduced in (4.7) and which defines the V-states equations. We shall only state some regularity results that fit with the assumptions of Crandall–Rabinowitz's theorem. For the proofs we refer the reader to the paper [22]. To begin with, we introduce the function spaces that we shall use. We set,

$$X = C_{ar}^{2-\alpha}(\mathbb{T}) \times C_{ar}^{2-\alpha}(\mathbb{T}), \quad Y = H \times H, \tag{4.32}$$

with

$$C_{ar}^{2-\alpha}(\mathbb{T}) = \left\{ f \in C^{2-\alpha}(\mathbb{T}); f(w) = \sum_{n \geq 1} a_n \overline{w}^n, w \in \mathbb{T}, a_n \in \mathbb{R}, n \in \mathbb{N}^* \right\}$$

and

$$H = \left\{ g \in C^{1-\alpha}(\mathbb{T}); g(w) = \frac{i}{2} \sum_{n \geq 1} a_n (w^n - \overline{w}^n), w \in \mathbb{T}, a_n \in \mathbb{R}, n \in \mathbb{N} \right\}.$$

For $b \in (0, 1)$, let V denote the product $B_r \times B_r$, where B_r is the open ball of X with center 0 and radius $r = (1/4) \min\{b, 1-b\}$. We note that this choice is done in order to guarantee that $\phi_1 = \mathrm{Id} + f_1$ and $\phi_2 = b\,\mathrm{Id} + f_2$ are conformal for $f_1, f_2 \in B_r$ and to prevent the intersection between the curves $\phi_1(\mathbb{T})$ and $\phi_2(\mathbb{T})$ which represent the boundaries of the V-states.

Now recall from (4.7) the form of the functional $G = (G_1, G_2)$,

$$G_j(\Omega, f_1, f_2) \triangleq \mathrm{Im}\left\{ \left(\Omega\,\phi_j(w) + S(\phi_2, \phi_j)(w) - S(\phi_1, \phi_j)(w) \right) \overline{w\,\phi_j'(w)} \right\},$$

$$w \in \mathbb{T}, \quad j = 1, 2,$$

where S is defined by

$$S(\phi_i, \phi_j)(w) = C_\alpha \fint_{\mathbb{T}} \frac{\phi_i'(\tau)\,d\tau}{|\phi_j(w) - \phi_i(\tau)|^\alpha}, \quad i, j = 1, 2. \tag{4.33}$$

We shall rewrite G_j as follows,

$$G_j(\Omega, f_1, f_2) = L_j(\Omega, f_j) + N_j(f_1, f_2), \quad j = 1, 2,$$

with

$$L_j(\Omega, f_j) \triangleq \mathrm{Im}\left\{ \left(\Omega\,\phi_j(w) + (-1)^j S(\phi_j, \phi_j)(w) \right) \overline{w}\,\overline{\phi_j'(w)} \right\},$$

and

$$N_j(f_1, f_2) \triangleq (-1)^{j-1} \mathrm{Im}\left\{ S(\phi_i, \phi_j)(w) \overline{w}\,\overline{\phi_j'(w)} \right\}, \quad i \neq j,$$

usually with the notation $\phi_1 = \mathrm{Id} + f_1$, $\phi_2 = b\,\mathrm{Id} + f_2$.

The main result reads as follows.

Proposition 4.4 The following holds true.

1. $G : \mathbb{R} \times V \to Y$ is well-defined.
2. $G : \mathbb{R} \times V \to Y$ is of class C^1.
3. The partial derivative $\partial_\Omega DG : \mathbb{R} \times V \to \mathcal{L}(X, Y)$ exists and is continuous.

4.5 Spectral Study

The main goal of this section is to perform a spectral study of the linearized operator of G at the annular solution $(\Omega, 0, 0)$ and denoted by the differential $DG(\Omega, 0, 0)$. In particular, we shall identify the values Ω for which the kernel of $DG(\Omega, 0, 0)$ is not trivial leading to what we call the dispersion relation. Therefore the next step is to look among the "nonlinear eigenvalues" Ω those corresponding to one-dimensional kernels which is an important assumption in Crandall–Rabinowitz's theorem. This task is very complicated compared to the previous cases discussed in [16, 19] due to the multiple parameters α, b, and m in this problem and especially to the nonlinear and implicit structure of the coefficients appearing in the dispersion relation.

4.5.1 Linearized Operator

We propose to compute explicitly the differential $DG(\Omega, 0, 0)$ and show that it acts as a matrix Fourier multiplier. Since $G = (G_1, G_2)$ then for given $(h_1, h_2) \in X$, we have

$$DG(\Omega, 0, 0)(h_1, h_2) = \begin{pmatrix} D_{f_1} G_1(\Omega, 0, 0) h_1 + D_{f_2} G_1(\Omega, 0, 0) h_2 \\ D_{f_1} G_2(\Omega, 0, 0) h_1 + D_{f_2} G_2(\Omega, 0, 0) h_2 \end{pmatrix},$$

where we recall the function spaces

$$X = C_{ar}^{2-\alpha}(\mathbb{T}) \times C_{ar}^{2-\alpha}(\mathbb{T}),$$

and

$$C_{ar}^{2-\alpha}(\mathbb{T}) = \left\{ f \in C^{2-\alpha}(\mathbb{T}); f(w) = \sum_{n \geq 1} a_n \overline{w}^n, w \in \mathbb{T}, a_n \in \mathbb{R}, n \in \mathbb{N}^\star \right\}.$$

Straightforward computations yield,

$$DG_1(\Omega, 0, 0)h(w) = \Omega \, \mathcal{L}_0(h_1)(w) - C_\alpha \, \mathcal{L}_1(h_1)(w) + C_\alpha \, \mathcal{L}_2(h_1, h_2)(w), \tag{4.34}$$

$$DG_2(\Omega, 0, 0)h(w) = b\Big(\Omega \, \mathcal{L}_0(h_2)(w) + b^{-\alpha} C_\alpha \, \mathcal{L}_1(h_2)(w) - C_\alpha \, \mathcal{L}_3(h_1, h_2)(w)\Big), \tag{4.35}$$

with

$$\mathcal{L}_0(h_j)(w) \triangleq \operatorname{Im}\Big\{ \overline{h_j'(w)} + \overline{w} h_j(w) \Big\},$$

$$\mathcal{L}_1(h_j)(w) \triangleq \operatorname{Im}\Big\{ \overline{w h_j'(w)} \fint_{\mathbb{T}} \frac{d\tau}{|w - \tau|^\alpha} + \overline{w} \fint_{\mathbb{T}} \frac{h_j'(\tau)}{|w - \tau|^\alpha} d\tau$$
$$- \alpha \overline{w} \fint_{\mathbb{T}} \frac{\operatorname{Re}\big[(w - \tau)\big(\overline{h_j(w)} - \overline{h_j(\tau)}\big)\big]}{|w - \tau|^{\alpha+2}} d\tau \Big\},$$

$$\mathcal{L}_2(h_1, h_2)(w) \triangleq \operatorname{Im}\Big\{ b\overline{w h_1'(w)} \fint_{\mathbb{T}} \frac{d\tau}{|w - b\tau|^\alpha} + \overline{w} \fint_{\mathbb{T}} \frac{h_2'(\tau) d\tau}{|w - b\tau|^\alpha}$$
$$- \alpha b\overline{w} \fint_{\mathbb{T}} \frac{\operatorname{Re}\big[(w - b\tau)\big(\overline{h_1(w)} - \overline{h_2(\tau)}\big)\big] d\tau}{|w - b\tau|^{\alpha+2}} \Big\},$$

and

$$\mathcal{L}_3(h_1, h_2)(w) \triangleq \operatorname{Im}\Big\{ b\overline{w} \, \overline{h_2'(w)} \fint_{\mathbb{T}} \frac{d\tau}{|bw - \tau|^\alpha} + \overline{w} \fint_{\mathbb{T}} \frac{h_1'(\tau)}{|bw - \tau|^\alpha} d\tau$$
$$- \alpha \overline{w} \fint_{\mathbb{T}} \frac{\operatorname{Re}\big[(bw - \tau)\big(\overline{h_2(w)} - \overline{h_1(\tau)}\big)\big]}{|bw - \tau|^{\alpha+2}} d\tau \Big\}.$$

We shall now compute the Fourier series of the mapping $w \mapsto DG(\Omega, 0, 0)(h_1, h_2)(w)$ with

$$h_1(w) = \sum_{n=1}^{\infty} a_n \overline{w}^n \quad \text{and} \quad h_2(w) = \sum_{n=1}^{\infty} c_n \overline{w}^n, \quad w \in \mathbb{T},$$

where a_n and c_n are real for all $n \in \mathbb{N}^\star$. This is summarized in the following lemma.

Lemma 4.5 Let $\alpha \in (0, 1)$ and $b \in (0, 1)$. We set

$$\Lambda_n(b) \triangleq \frac{1}{b} \int_0^{+\infty} \frac{J_n(bt) J_n(t)}{t^{1-\alpha}} dt, \tag{4.36}$$

and

$$\Theta_n \triangleq \Lambda_1(1) - \Lambda_n(1),$$

where J_n refers to the Bessel function of the first kind. Then, we have

$$DG(\Omega, 0, 0)(h_1, h_2)(w) = \frac{i}{2} \sum_{n \geq 1} (n+1) M_{n+1}^\alpha \begin{pmatrix} a_n \\ c_n \end{pmatrix} \left(w^{n+1} - \overline{w}^{n+1} \right),$$

(4.37)

where the matrix M_n is given for $n \geq 2$ by

$$M_n^\alpha \triangleq \begin{pmatrix} \Omega - \Theta_n + b^2 \Lambda_1(b) & -b^2 \Lambda_n(b) \\ b\Lambda_n(b) & b\Omega + b^{1-\alpha}\Theta_n - b\Lambda_1(b) \end{pmatrix}.$$

(4.38)

The determinant of this matrix is given by

$$\det\left(M_n^\alpha\right) = \left(\Omega - \Theta_n + b^2 \Lambda_1(b)\right)\left(b\Omega + b^{1-\alpha}\Theta_n - b\Lambda_1(b)\right) + b^3 \Lambda_n^2(b).$$

(4.39)

Proof First we shall compute $DG_1(\Omega, 0, 0)(h_1, h_2)$. For this goal we start with calculating the term $\mathcal{L}_0(h_1(w))$ on the right-hand side of (4.34) which is easy compared to the other terms. Thus by straightforward computations we obtain

$$\mathcal{L}_0(h_1)(w) = \text{Im}\left\{ \sum_{n \geq 1} \left(a_n \overline{w}^{n+1} - n a_n w^{n+1} \right) \right\}$$

$$= \frac{i}{2} \sum_{n \geq 1} (n+1) a_n \left(w^{n+1} - \overline{w}^{n+1} \right).$$

(4.40)

The computation of the second term $\mathcal{L}_1(h_1)(w)$ was done in the paper [16] dealing with the simply connected domain. It is given by

$$C_\alpha \mathcal{L}_1(h_1)(w) = \frac{i}{2} \sum_{n \geq 1} a_n(n+1) \frac{\Gamma(1-\alpha)}{2^{1-\alpha}\Gamma^2(1-\frac{\alpha}{2})}$$

$$\times \left(\frac{\Gamma(1+\frac{\alpha}{2})}{\Gamma(2-\frac{\alpha}{2})} - \frac{\Gamma(n+1+\frac{\alpha}{2})}{\Gamma(n+2-\frac{\alpha}{2})} \right) \left(w^{n+1} - \overline{w}^{n+1} \right).$$

Manipulating some algebraic identities we find that,

$$C_\alpha \mathcal{L}_1(h_1)(w) = \frac{i}{2} \sum_{n \geq 1} a_n(n+1) \Theta_{n+1} \left(w^{n+1} - \overline{w}^{n+1} \right).$$

(4.41)

To compute the term $\mathcal{L}_2(h_1, h_2)(w)$ we first split it into two parts as follows,

$$\mathcal{L}_2(h_1, h_2)(w) = \text{Im}\left\{ I_1(w) + I_2(w) \right\},$$

(4.42)

with

$$I_1(w) \triangleq \frac{b\overline{h_1'(w)}}{w} \fint_{\mathbb{T}} \frac{d\tau}{|w - b\tau|^\alpha} - \frac{\alpha b}{2w} \fint_{\mathbb{T}} \frac{(w - b\tau)\big(\overline{h_1(w)} - \overline{h_2(\tau)}\big)}{|w - b\tau|^{\alpha+2}} d\tau$$

and

$$I_2(w) \triangleq \frac{1}{w} \fint_{\mathbb{T}} \frac{h_2'(\tau)}{|w - b\tau|^\alpha} d\tau - \frac{\alpha b}{2w} \fint_{\mathbb{T}} \frac{(\overline{w} - b\overline{\tau})\big(h_1(w) - h_2(\tau)\big)}{|w - b\tau|^{\alpha+2}} d\tau.$$

By using the Fourier expansions of h_1 and h_2 we get,

$$I_1(w) = -b \sum_{n \geq 1} n a_n w^n \fint_{\mathbb{T}} \frac{d\tau}{|w - b\tau|^\alpha} - \frac{\alpha}{2} b\overline{w} \sum_{n \geq 1} \fint_{\mathbb{T}} \frac{(w - b\tau)(a_n w^n - c_n \tau^n)}{|w - b\tau|^{\alpha+2}} d\tau.$$

Then by applying the formula (4.27) with $n = 1$ to the first term and the formula (4.28) to the second term we find

$$I_1(w) = -\frac{\alpha}{2} b^2 \sum_{n \geq 1} n a_n w^{n+1} F\Big(\frac{\alpha}{2}, 1 + \frac{\alpha}{2}; 2; b^2\Big)$$

$$- \frac{\alpha}{2} b^2 \sum_{n \geq 0} w^{n+1} \Bigg[a_n \Big(1 + \frac{\alpha}{2}\Big) F\Big(\frac{\alpha}{2}, 2 + \frac{\alpha}{2}; 2; b^2\Big)$$

$$- c_n b^n \frac{(1 + \frac{\alpha}{2})_{n+1}}{(n+1)!} F\Big(\frac{\alpha}{2}, n + 2 + \frac{\alpha}{2}; n + 2; b^2\Big) \Bigg]$$

$$\triangleq -\frac{\alpha}{2} b^2 \sum_{n \geq 0} \big(a_n \gamma_n + c_n \delta_n \big) w^{n+1}, \qquad (4.43)$$

where we have used in the last equality the notation,

$$\gamma_n \triangleq \Big(1 + \frac{\alpha}{2}\Big) F\Big(\frac{\alpha}{2}, 2 + \frac{\alpha}{2}; 2; b^2\Big) + n F\Big(\frac{\alpha}{2}, 1 + \frac{\alpha}{2}; 2; b^2\Big)$$

and

$$\delta_n = -b^n \frac{(1 + \frac{\alpha}{2})_{n+1}}{(n+1)!} F\Big(\frac{\alpha}{2}, n + 2 + \frac{\alpha}{2}; n + 2; b^2\Big).$$

Similarly, the second term $I_2(w)$ may be written in the form,

$$I_2(w) = -\overline{w} \sum_{n \geq 0} n c_n \fint_{\mathbb{T}} \frac{\overline{\tau}^{n+1} d\tau}{|w - b\tau|^\alpha} - \frac{\alpha}{2} b\overline{w} \sum_{n \geq 0} \fint_{\mathbb{T}} \frac{(\overline{w} - b\overline{\tau})(a_n \overline{w}^n - c_n \overline{\tau}^n)}{|w - b\tau|^{\alpha+2}} d\tau.$$

Using the elementary fact

$$\fint_{\mathbb{T}} \frac{\overline{\tau}^{n+1} d\tau}{|w - b\tau|^\alpha} = \overline{\fint_{\mathbb{T}} \frac{\tau^{n-1} d\tau}{|w - b\tau|^\alpha}} \qquad (4.44)$$

combined with the formulae (4.27) and (4.29) we obtain

$$
\begin{aligned}
I_2(w) = &- \sum_{n\geq 1} nc_n \frac{(\frac{\alpha}{2})_n}{n!} \overline{w}^{n+1} b^n F\left(\frac{\alpha}{2}, n + \frac{\alpha}{2}; n + 1; b^2\right) \\
&- \frac{\alpha}{2} \sum_{n\geq 1} \overline{w}^{n+1} \left[a_n b^2 \frac{\alpha}{2} F\left(\frac{\alpha}{2} + 1, \frac{\alpha}{2} + 1; 2; b^2\right) \right.\\
&\left. - c_n b^n \frac{(1 + \frac{\alpha}{2})_{n-1}}{(n-1)!} F\left(\frac{\alpha}{2}, n + \frac{\alpha}{2}; n; b^2\right) \right] \\
= &- \frac{\alpha}{2} \sum_{n\geq 1} \overline{w}^{n+1} \left[a_n b^2 \frac{\alpha}{2} F(\frac{\alpha}{2} + 1, \frac{\alpha}{2} + 1; 2; b^2) \right.\\
&+ c_n \frac{(1 + \frac{\alpha}{2})_{n-1}}{(n-1)!} b^n \left(F\left(\frac{\alpha}{2}, n + \frac{\alpha}{2}; n + 1; b^2\right) \right.\\
&\left.\left. - F\left(\frac{\alpha}{2}, n + \frac{\alpha}{2}; n; b^2\right) \right) \right].
\end{aligned}
\tag{4.45}
$$

Thus owing to the formula (4.17) applied with $a = \frac{\alpha}{2}, b = \frac{\alpha}{2} + n$ and $c = n$ one gets

$$
I_2(w) = -\frac{\alpha}{2} b^2 \sum_{n\geq 1} \left(\alpha_n a_n + \beta_n c_n \right) \overline{w}^{n+1},
\tag{4.46}
$$

where α_n and β_n are defined by,

$$
\alpha_n \triangleq \frac{\alpha}{2} F\left(\frac{\alpha}{2} + 1, \frac{\alpha}{2} + 1; 2; b^2\right)
$$

and

$$
\beta_n \triangleq -\frac{(\frac{\alpha}{2})_{n+1}}{(n+1)!} b^n F\left(\frac{\alpha}{2} + 1, n + 1 + \frac{\alpha}{2}; n + 2; b^2\right).
$$

Inserting the identities (4.46) and (4.43) into (4.42) we find

$$
\begin{aligned}
\mathcal{L}_2(h_1, h_2)(w) &= -\frac{\alpha}{2} b^2 \mathrm{Im} \left\{ \sum_{n\geq 1} \left(a_n \gamma_n + c_n \delta_n \right) w^{n+1} + \sum_{n\geq 1} \left(a_n \alpha_n + c_n \beta_n \right) \overline{w}^{n+1} \right\} \\
&= i \frac{\alpha}{4} b^2 \sum_{n\geq 1} \left(w^{n+1} - \overline{w}^{n+1} \right) \left[a_n (\gamma_n - \alpha_n) + c_n (\delta_n - \beta_n) \right].
\end{aligned}
$$

To compute $\gamma_n - \alpha_n$ we shall use the formula (4.22) which gives

$$\gamma_n - \alpha_n = \left(1 + \frac{\alpha}{2}\right) F\left(\frac{\alpha}{2}, 2 + \frac{\alpha}{2}; 2; b^2\right) - \frac{\alpha}{2} F\left(\frac{\alpha}{2} + 1, \frac{\alpha}{2} + 1; 2; b^2\right)$$
$$+ n F\left(\frac{\alpha}{2}, 1 + \frac{\alpha}{2}; 2; b^2\right)$$
$$= (n + 1) F\left(\frac{\alpha}{2}, 1 + \frac{\alpha}{2}; 2; b^2\right).$$

Similarly we have

$$\delta_n - \beta_n = -b^n \frac{(1 + \frac{\alpha}{2})_n}{(n+1)!} \left[\left(n + 1 + \frac{\alpha}{2}\right) F\left(\frac{\alpha}{2}, n + 2 + \frac{\alpha}{2}; n + 2; b^2\right) \right.$$
$$\left. - \frac{\alpha}{2} F\left(\frac{\alpha}{2} + 1, n + 1 + \frac{\alpha}{2}; n + 2; b^2\right) \right]$$

and therefore using once again the identity (4.22) we find

$$\delta_n - \beta_n = -b^n \frac{(1 + \frac{\alpha}{2})_n}{n!} F\left(\frac{\alpha}{2}, n + 1 + \frac{\alpha}{2}, n + 2, b^2\right).$$

Consequently the Fourier expansion of $\mathcal{L}_2(h_1, h_2)$ is described by the formula

$$\mathcal{L}_2(h_1, h_2)(w) = \frac{i}{2} b^2 \sum_{n \geq 1} (n + 1) \left[a_n \frac{\alpha}{2} F\left(\frac{\alpha}{2}, 1 + \frac{\alpha}{2}, 2, b^2\right) \right.$$
$$\left. - c_n b^n \frac{(\frac{\alpha}{2})_{n+1}}{(n+1)!} F\left(\frac{\alpha}{2}, n + 1 + \frac{\alpha}{2}, n + 2, b^2\right) \right] \left(w^{n+1} - \overline{w}^{n+1} \right).$$

By virtue of the identity (4.25) we get

$$C_\alpha \mathcal{L}_2(h_1, h_2)(w) = \frac{i}{2} \sum_{n \geq 1} (n + 1) \left[a_n b^2 \Lambda_1(b) - c_n b^2 \Lambda_{n+1}(b) \right] \left(w^{n+1} - \overline{w}^{n+1} \right).$$

$$(4.47)$$

Finally inserting (4.40), (4.41), and (4.47) into (4.34) we find

$$DG_1(\Omega, 0, 0)(h_1, h_2)(w) = \frac{i}{2} \sum_{n \geq 1} (n + 1) \left[a_n \left(\Omega - \Theta_{n+1} + b^2 \Lambda_1(b) \right) \right.$$
$$\left. - c_n b^2 \Lambda_{n+1}(b) \right] \left(w^{n+1} - \overline{w}^{n+1} \right).$$

$$(4.48)$$

Next, we shall move to the computations of $DG_2(\Omega, 0, 0)(h_1, h_2)$ defined in (4.35). The first two terms are done in the preceding step and therefore it remains just to compute the term $\mathcal{L}_3(h_1, h_2)$. It may be split into two terms,

$$\mathcal{L}_3(h_1, h_2)(w) = \mathrm{Im}\left\{\tilde{I}_1(w) + \tilde{I}_2(w)\right\}, \qquad (4.49)$$

with

$$\tilde{I}_1(w) \triangleq \frac{\overline{h_2'(w)}}{bw} \fint_{\mathbb{T}} \frac{d\tau}{|bw - \tau|^\alpha} - \frac{\alpha}{2w} \fint_{\mathbb{T}} \frac{(bw - \tau)\left(\overline{h_2(w)} - \overline{h_1(\tau)}\right)}{|bw - \tau|^{\alpha+2}} d\tau$$

and

$$\tilde{I}_2(w) \triangleq \frac{1}{w} \fint_{\mathbb{T}} \frac{h_1'(\tau) d\tau}{|bw - \tau|^\alpha} - \frac{\alpha}{2w} \fint_{\mathbb{T}} \frac{(b\overline{w} - \overline{\tau})\left(h_2(w) - h_1(\tau)\right)}{|bw - \tau|^{\alpha+2}} d\tau.$$

To compute the first term $\tilde{I}_1(w)$ we write

$$\tilde{I}_1(w) = -\sum_{n\geq 1} n\frac{c_n}{b} w^n \fint_{\mathbb{T}} \frac{d\tau}{|bw - \tau|^\alpha} - \frac{\alpha}{2w} \sum_{n\geq 1} \fint_{\mathbb{T}} \frac{(bw - \tau)\left(c_n w^n - a_n \tau^n\right)}{|bw - \tau|^{\alpha+2}} d\tau.$$

Thus applying successively the formula (4.27) to the first term with $n = 1$ and the formula (4.30) to the second one we get

$$\tilde{I}_1(w) = -\frac{\alpha}{2} \sum_{n\geq 1} n\, c_n \; w^{n+1} F\left(\frac{\alpha}{2}, 1 + \frac{\alpha}{2}; 2; b^2\right)$$

$$+ \frac{\alpha}{2} \sum_{n\geq 1} w^{n+1} \left[c_n b^2 \frac{\alpha}{4}\left(1 + \frac{\alpha}{2}\right) F\left(1 + \frac{\alpha}{2}, 2 + \frac{\alpha}{2}; 3; b^2\right) \right.$$

$$\left. - a_n b^{n+2} \frac{(\frac{\alpha}{2})_{n+2}}{(n+2)!} F\left(1 + \frac{\alpha}{2}, n + 2 + \frac{\alpha}{2}; n + 3; b^2\right) \right]$$

$$= -\frac{\alpha}{2} \sum_{n\geq 1} \left(a_n\, \tilde{\gamma}_n + c_n\, \tilde{\delta}_n \right) w^{n+1} \qquad (4.50)$$

with

$$\tilde{\gamma}_n \triangleq b^{n+2} \frac{(\frac{\alpha}{2})_{n+2}}{(n+2)!} F\left(1 + \frac{\alpha}{2}, n + 2 + \frac{\alpha}{2}; n + 3; b^2\right)$$

and

$$\tilde{\delta}_n \triangleq n F\left(\frac{\alpha}{2}, 1 + \frac{\alpha}{2}; 2; b^2\right) - b^2 \frac{\alpha}{4}\left(1 + \frac{\alpha}{2}\right) F\left(1 + \frac{\alpha}{2}, 2 + \frac{\alpha}{2}; 3; b^2\right).$$

As to the term $\tilde{I}_2(w)$ we write

$$\tilde{I}_2(w) = -\frac{1}{w} \sum_{n\geq 1} n a_n \fint_{\mathbb{T}} \frac{\overline{\tau}^{n+1} d\tau}{|bw - \tau|^\alpha} - \frac{\alpha}{2w} \sum_{n\geq 1} \fint_{\mathbb{T}} \frac{(b\overline{w} - \overline{\tau})\left(c_n \overline{w}^n - a_n \overline{\tau}^n\right) d\tau}{|bw - \tau|^{\alpha+2}}.$$

Owing to (4.44) and using the formulae (4.27) and (4.31), one gets

$$
\begin{aligned}
\tilde{I}_2(w) = & -\sum_{n\geq 1} n a_n \overline{w}^{n+1} b^n \frac{(\frac{\alpha}{2})_n}{n!} F\left(\frac{\alpha}{2}, n+\frac{\alpha}{2}; n+1; b^2\right) \\
& -\frac{\alpha}{2}\sum_{n\geq 1}\overline{w}^{n+1}\left[-c_n F\left(\frac{\alpha}{2},\frac{\alpha}{2}+1;1;b^2\right)\right. \\
& \left. + a_n b^n \frac{(\frac{\alpha}{2})_n}{n!} F\left(\frac{\alpha}{2}+1, n+\frac{\alpha}{2}; n+1; b^2\right)\right] \\
= & -\frac{\alpha}{2}\sum_{n\geq 1}\left(a_n\tilde{\alpha}_n + c_n\tilde{\beta}_n\right)\overline{w}^{n+1},
\end{aligned} \tag{4.51}
$$

with

$$
\tilde{\alpha}_n \triangleq \frac{(1+\frac{\alpha}{2})_{n-1}}{n!} b^n\left[n F\left(\frac{\alpha}{2}, n+\frac{\alpha}{2}; n+1; b^2\right)+\frac{\alpha}{2}F\left(1+\frac{\alpha}{2}, n+\frac{\alpha}{2}; n+1; b^2\right)\right]
$$

and

$$
\tilde{\beta}_n \triangleq -F\left(\frac{\alpha}{2},\frac{\alpha}{2}+1;1;b^2\right).
$$

Now inserting the identities (4.50) and (4.51) into (4.49) we find

$$
\begin{aligned}
\mathcal{L}_3(h_1,h_2)(w) & = -\frac{\alpha}{2}\mathrm{Im}\left\{\sum_{n\geq 1}\left(a_n\tilde{\gamma}_n + c_n\tilde{\delta}_n\right)w^{n+1} + \sum_{n\geq 1}\left(a_n\tilde{\alpha}_n + c_n\tilde{\beta}_n\right)\overline{w}^{n+1}\right\} \\
& = i\frac{\alpha}{4}\sum_{n\geq 1}\left[a_n\left(\tilde{\gamma}_n - \tilde{\alpha}_n\right) + c_n\left(\tilde{\delta}_n - \tilde{\beta}_n\right)\right]\left(w^{n+1} - \overline{w}^{n+1}\right). \tag{4.52}
\end{aligned}
$$

From the foregoing expressions for $\tilde{\gamma}_n$ and $\tilde{\alpha}_n$ one may write,

$$
\begin{aligned}
\tilde{\gamma}_n - \tilde{\alpha}_n = & \; b^{n+2}\frac{(\frac{\alpha}{2})_{n+2}}{(n+2)!} F\left(1+\frac{\alpha}{2}, n+2+\frac{\alpha}{2}; n+3; b^2\right) \\
& -\frac{(1+\frac{\alpha}{2})_{n-1}}{n!} b^n\left[n F\left(\frac{\alpha}{2}, n+\frac{\alpha}{2}; n+1; b^2\right)+\frac{\alpha}{2}F\left(1+\frac{\alpha}{2}, n+\frac{\alpha}{2}; n+1; b^2\right)\right] \\
= & \;\frac{(1+\frac{\alpha}{2})_{n-1}}{n!} b^n\left[b^2 \frac{\frac{\alpha}{2}(\frac{\alpha}{2}+n)(\frac{\alpha}{2}+1+n)}{(n+2)(n+1)} F\left(1+\frac{\alpha}{2}, n+2+\frac{\alpha}{2}; n+3; b^2\right)\right. \\
& \left. -n F\left(\frac{\alpha}{2}, n+\frac{\alpha}{2}; n+1; b^2\right)-\frac{\alpha}{2}F\left(1+\frac{\alpha}{2}, n+\frac{\alpha}{2}; n+1; b^2\right)\right].
\end{aligned}
$$

Hence using the formula (4.17) with $a = \frac{\alpha}{2}, b = n + 1 + \frac{\alpha}{2}$ and $c = n + 1$ yields

$$\tilde{\gamma}_n - \tilde{\alpha}_n = \frac{(1 + \frac{\alpha}{2})_{n-1}}{n!} b^n \left[\left(\frac{\alpha}{2} + n \right) F\left(\frac{\alpha}{2}, n + 1 + \frac{\alpha}{2}; n + 1; b^2 \right) \right.$$
$$- \left(\frac{\alpha}{2} + n \right) F\left(\frac{\alpha}{2}, n + 1 + \frac{\alpha}{2}; n + 2; b^2 \right)$$
$$- nF\left(\frac{\alpha}{2}, n + \frac{\alpha}{2}; n + 1; b^2 \right)$$
$$\left. - \frac{\alpha}{2} F\left(1 + \frac{\alpha}{2}, n + \frac{\alpha}{2}; n + 1; b^2 \right) \right].$$

Applying the formula (4.22) with $a = \frac{\alpha}{2}, b = \frac{\alpha}{2} + n$ and $c = n + 1$ we get

$$\left(\frac{\alpha}{2} + n \right) F\left(\frac{\alpha}{2}, n + 1 + \frac{\alpha}{2}; n + 1; b^2 \right) - \frac{\alpha}{2} F\left(1 + \frac{\alpha}{2}, n + \frac{\alpha}{2}; n + 1; b^2 \right)$$
$$= nF\left(\frac{\alpha}{2}, n + \frac{\alpha}{2}; n + 1; b^2 \right).$$

This implies,

$$\tilde{\gamma}_n - \tilde{\alpha}_n = -\frac{(1 + \frac{\alpha}{2})_n}{n!} b^n F\left(\frac{\alpha}{2}, n + 1 + \frac{\alpha}{2}; n + 2; b^2 \right).$$

Using the expressions of $\tilde{\delta}_n$ and $\tilde{\beta}_n$ combined with the identity (4.17) applied with $a = \frac{\alpha}{2}, b = 1 + \frac{\alpha}{2}$ and $c = 1$ we find the compact formula

$$\tilde{\delta}_n - \tilde{\beta}_n = nF\left(\frac{\alpha}{2}, 1 + \frac{\alpha}{2}; 2; b^2 \right) - b^2 \frac{\alpha}{4} \left(1 + \frac{\alpha}{2} \right) F\left(1 + \frac{\alpha}{2}, 2 + \frac{\alpha}{2}; 3; b^2 \right)$$
$$+ F\left(\frac{\alpha}{2}, \frac{\alpha}{2} + 1; 1; b^2 \right)$$
$$= (n + 1)F\left(\frac{\alpha}{2}, 1 + \frac{\alpha}{2}; 2; b^2 \right).$$

Putting together the preceding identities allows to write

$$\mathcal{L}_3(h_1, h_2(w)) = i\frac{\alpha}{4} \sum_{n \geq 1} (n + 1) \left(w^{n+1} - \overline{w}^{n+1} \right) \left[c_n F\left(\frac{\alpha}{2}, 1 + \frac{\alpha}{2}; 2; b^2 \right) \right.$$
$$\left. - a_n b^n \frac{(1 + \frac{\alpha}{2})_n}{(n + 1)!} F\left(\frac{\alpha}{2}, n + 1 + \frac{\alpha}{2}; n + 2, b^2 \right) \right].$$

According to the identity (4.25) we get

$$C_\alpha \mathcal{L}_3(h_1, h_2(w)) = \frac{i}{2} \sum_{n \geq 1} (n + 1) \left(c_n \Lambda_1(b) - a_n \Lambda_{n+1}(b) \right) \left(w^{n+1} - \overline{w}^{n+1} \right).$$

Finally, inserting the preceding identity and the expressions (4.41) and (4.49) into (4.35) one can readily verify that

$$DG_2(\Omega,0,0)(h_1,h_2)(w) = \frac{i}{2} \sum_{n\geq 1} (n+1) \Big[a_n b \Lambda_{n+1}(b)$$

$$+ c_n \Big(b\Omega + b^{1-\alpha}\Theta_{n+1} - b\Lambda_1(b) \Big) \Big] \times \Big(w^{n+1} - \overline{w}^{n+1} \Big).$$

$$(4.53)$$

This concludes the proof of the Lemma 4.5.   □

4.5.2 Monotonicity of the Eigenvalues

In this section we shall discuss some important properties concerning the monotonicity of the eigenvalues associated with the matrix M_n^α already seen in Lemma 4.5. This will be crucial in the study of the kernel of the linearized operator $DG(\Omega,0,0)$. Recall that

$$M_n^\alpha = \begin{pmatrix} \Omega - \Theta_n + b^2\Lambda_1(b) & -b^2\Lambda_n(b) \\ b\Lambda_n(b) & b\Omega + b^{1-\alpha}\Theta_n - b\Lambda_1(b) \end{pmatrix}.$$

The determinant of this matrix given by (4.39) is a second-order polynomial on the variable Ω and therefore it has two roots depending on all the parameters n, b, and α. For our deal it is important to formulate sufficient conditions to avoid the eigenvalues crossing in order to guarantee a one-dimensional kernel which is an essential assumption in Crandall–Rabinowitz's theorem. In what follows we shall use the variable $\lambda \triangleq 1 - 2\Omega$ instead of Ω in the spirit of the work of [19]. Thus easy computations show that the determinant (4.39) takes the form,

$$\det\left(M_n^\alpha\right) = b\left(\lambda^2 - 2C_n\lambda + D_n\right),$$

$$(4.54)$$

with

$$C_n \triangleq 1 + \left(b^{-\alpha} - 1\right)\Theta_n - \left(1 - b^2\right)\Lambda_1(b),$$

and

$$D_n \triangleq -4b^{-\alpha}\Theta_n^2 + 2\left[b^{-\alpha} - 1 + 2\left(1 + b^{2-\alpha}\right)\Lambda_1(b)\right]\Theta_n$$

$$- 4b^2\left(\Lambda_1^2(b) - \Lambda_n^2(b)\right) - 2\left(1 - b^2\right)\Lambda_1(b) + 1.$$

$$(4.55)$$

Note that the quantities $\Lambda_n(b)$ and Θ_n have been introduced in Lemma 4.5. It is easy to check through straightforward computations that the reduced discriminant of the second order polynomial appearing in (4.54) is given by

$$\Delta_n = \left((b^{-\alpha} + 1)\Theta_n - (1 + b^2)\Lambda_1(b) \right)^2 - 4b^2\Lambda_n^2(b). \qquad (4.56)$$

Our result reads as follows and the proof is given in [22].

Proposition 4.6 Let $N \geq 2$ be the smallest integer satisfying,

$$\Theta_N > \frac{(1 + b^2)}{b^{-\alpha} + 1}\Lambda_1(b) + \frac{2b}{b^{-\alpha} + 1}\Lambda_N(b).$$

Then the following holds true.

1. For all $n \geq N$ we get $\Delta_n > 0$ and (4.54) admits two different real solutions given by

$$\lambda_n^{\pm} \triangleq C_n \pm \sqrt{\Delta_n}.$$

2. The sequences $(\Delta_n)_{n \geq N}$ and $(\lambda_n^+)_{n \geq N}$ are strictly increasing and $(\lambda_n^-)_{n \geq N}$ is strictly decreasing.
3. For all $m > n \geq N$ we have

$$\lambda_m^- < \lambda_n^- < \lambda_n^+ < \lambda_m^+.$$

4.6 Bifurcation at Simple Eigenvalues

We shall prove Theorem 4.1 which is deeply related to the spectral study developed in the preceding section combined with Crandall–Rabinowitz's theorem. To construct the function spaces where the bifurcation occurs we shall take into account the restriction to the high frequencies stated in Proposition 4.6 and include the m-fold symmetry of the V-states. To proceed, fix $b \in (0, 1)$ and $m \geq N$, where N is defined in Proposition 4.6. Set

$$X_m = C_m^{2-\alpha}(\mathbb{T}) \times C_m^{2-\alpha}(\mathbb{T}),$$

where $C_m^{2-\alpha}(\mathbb{T})$ is the space of the 2π−periodic functions $f \in C^{2-\alpha}(\mathbb{T})$ whose Fourier series is given by

$$f(w) = \sum_{n=1}^{\infty} a_n \overline{w}^{nm-1}, \quad w \in \mathbb{T}, \quad a_n \in \mathbb{R}.$$

This space is equipped with its usual norm. We define the ball of radius $r \in (0,1)$ by

$$B_r^m = \left\{ f \in X_m, \ \|f\|_{C^{2-\alpha}(\mathbb{T})} \leq r \right\}$$

and we introduce the neighborhood of zero,

$$V_{m,r} \triangleq B_r^m \times B_r^m.$$

The set $V_{m,r}$ is endowed with the induced topology of the product spaces. Take $(f_1, f_2) \in V_{m,r}$ then the expansions of the associated conformal mappings ϕ_1, ϕ_2 outside the unit disk $\{z \in \mathbb{C}; \ |z| \geq 1\}$ are given successively by

$$\phi_1(z) = z + f_1(z) = z\left(1 + \sum_{n=1}^{\infty} \frac{a_n}{z^{nm}}\right)$$

and

$$\phi_2(z) = bz + f_2(z) = z\left(b + \sum_{n=1}^{\infty} \frac{b_n}{z^{nm}}\right).$$

This structure provides the m−fold symmetry of the associated boundaries $\phi_1(\mathbb{T})$ and $\phi_2(\mathbb{T})$, via the relation

$$\phi_j(e^{2i\pi/m}z) = e^{2i\pi/m}\phi_j(z), \quad j = 1, 2 \quad \text{and} \quad |z| \geq 1. \tag{4.57}$$

For functions f_1 and f_2 with small size the boundaries can be seen as a small perturbation of the boundaries of the annulus $\{z \in \mathbb{C}; \ b \leq |z| \leq 1\}$. Set

$$H_m = \left\{ g \in C^{1-\alpha}(\mathbb{T}), \ g(w) = i\sum_{n\geq1} A_n\left(w^{mn} - \overline{w}^{mn}\right), \ A_n \in \mathbb{R}, \ n \in \mathbb{N}^* \right\}$$

and define the product space Y_m by

$$Y_m = H_m \times H_m.$$

From Proposition 4.6 recall the definition of the eigenvalues $\lambda_m^\pm$ and the associated angular velocities are

$$\Omega_m^\pm = \frac{1}{2} - \frac{1}{2}\lambda_m^\pm$$
$$= \frac{1}{2}\widehat{C}_m \mp \frac{1}{2}\sqrt{\Delta_m}$$

with

$$\Delta_m \triangleq \left((b^{-\alpha} + 1)\Theta_m - (1 + b^2)\Lambda_1(b)\right)^2 - 4b^2\Lambda_m^2(b)$$

and

$$\widehat{C}_m \triangleq (1 - b^{-\alpha})\Theta_m + (1 - b^2)\Lambda_1(b).$$

Note that Θ_m and $\Lambda_m(b)$ were introduced in Lemma 4.5. The V-states equations are described in (4.6) and (4.7) which we restate here: for $j \in \{1,2\}$,

$$F_j(\Omega, \phi_1, \phi_2)(w) \triangleq G_j(\Omega, f_1, f_2)(w) = 0, \quad \forall w \in \mathbb{T}; \quad \text{and} \quad G \triangleq (G_1, G_2).$$

with

$$F_j(\Omega, \phi_1, \phi_2)(w) \triangleq \Omega \, \mathrm{Im}\left\{ \phi_j(w)\overline{w} \, \overline{\phi_j'(w)} \right\}$$

$$+ C_\alpha \, \mathrm{Im}\left\{ \left(\fint_{\mathbb{T}} \frac{\phi_2'(\tau)d\tau}{|\phi_j(w) - \phi_2(\tau)|^\alpha} - \fint_{\mathbb{T}} \frac{\phi_1'(\tau)d\tau}{|\phi_j(w) - \phi_1(\tau)|^\alpha} \right) \overline{w} \, \overline{\phi_j'(w)} \right\}.$$

Now, to apply Crandall–Rabinowitz's theorem it suffices to show the following result which is proved in [22].

Proposition 4.7 Let N be as in the part (1) of Proposition 4.6 and $m \geq N$, and take $\Omega \in \{\Omega_m^\pm\}$. Then, the following assertions hold true.

1. There exists $r > 0$ such that $G : \mathbb{R} \times V_{m,r} \to Y_m$ is well-defined and of class C^1.
2. The kernel of $DG(\Omega, 0, 0)$ is one-dimensional and generated by

$$v_{0,m} : w \in \mathbb{T} \mapsto \begin{pmatrix} \Omega + b^{-\alpha}\Theta_m - \Lambda_1(b) \\ -\Lambda_m(b) \end{pmatrix} \overline{w}^{m-1}.$$

3. The range of $DG(\Omega, 0, 0)$ is closed and is of co-dimension one in Y_m.
4. Transversality assumption: If Ω is a simple eigenvalue ($\Delta_m > 0$) then

$$\partial_\Omega DG(\Omega_m^\pm, 0, 0)v_{0,m} \notin R\big(DG(\Omega_m^\pm, 0, 0)\big).$$

References

[1] H. Aref, Integrable, chaotic, and turbulent vortex motion in two-dimensional flows. *Ann. Rev. Fluid Mech.*, 15 (1983), 345–89.

[2] A. L. Bertozzi, P. Constantin, Global regularity for vortex patches. *Comm. Math. Phys.* 152 (1993), no. 1, 19–28.

[3] A. Bertozzi, A. Majda, *Vorticity and Incompressible Flow*. Cambridge Texts in Applied Mathematics, Cambridge University Press, Cambridge (2002).

[4] J. Burbea, Motions of vortex patches. *Lett. Math. Phys.* 6 (1982), 1–16.

[5] A. Castro, D. Córdoba, J. Gómez-Serrano, A. Martín Zamora, Remarks on geometric properties of SQG sharp fronts and α-patches. *Discrete Contin. Dyn. Syst.* 34 (2014), no. 12, 5045–59.

[6] A. Castro, D. Córdoba, J. Gómez-Serrano. Existence and regularity of rotating global solutions for the generalized surface quasi-geostrophic equations. *Duke Math. J.* 165 (2016), no. 5, 935–84.

[7] D. Chae, P. Constantin, D. Córdoba, F. Gancedo, J. Wu, Generalized surface quasi-geostrophic equations with singular velocities. *Comm. Pure Appl. Math.* 65 (2012), no. 8, 1037–66.

[8] S. A. Chaplygin, On a pulsating cylindrical vortex. Translated from the 1899 Russian original by G. Krichevets, edited by D. Blackmore and with comments by V. V. Meleshko. *Regul. Chaotic Dyn.* 12 (2007), no. 1, 101–16.

[9] J.-Y. Chemin, Fluides Parfaits Incompressibles, Astérisque 230 (1995); *Perfect Incompressible Fluids*, translated by I. Gallagher and D. Iftimie, Oxford Lecture Series in Mathematics and Its Applications, Vol. 14, Clarendon Press–Oxford University Press, New York (1998).

[10] P. Constantin, A. J. Majda, E. Tabak, Formation of strong fronts in the 2-D quasigeostrophic thermal active scalar. *Nonlinearity* 7 (1994), no. 6, 1495–1533.

[11] D. Córdoba, M. A. Fontelos, A. M. Mancho, J. L. Rodrigo, Evidence of singularities for a family of contour dynamics equations. *Proc. Natl. Acad. Sci. USA* 102 (2005), no. 17, 5949–52.

[12] M. G. Crandall, P.H. Rabinowitz, Bifurcation from simple eigenvalues, *J. Funct. Anal.* 8 (1971), 321–40.

[13] G. S. Deem and N. J. Zabusky, Vortex waves: Stationary "V-states," interactions, recurrence, and breaking, *Phys. Rev. Lett.* 40 (1978), no. 13, 859–62.

[14] G. R. Flierl, L. M. Polvani, Generalized Kirchhoff vortices. *Phys. Fluids* 29 (1986), 2376–9.

[15] F. Gancedo, Existence for the α-patch model and the QG sharp front in Sobolev spaces. *Adv. Math.* 217 (2008), no. 6, 2569–98.

[16] Z. Hassainia, T. Hmidi. On the V-states for the generalized quasi-geostrophic equations. *Comm. Math. Phys.* 337 (2015), no. 1, 321–77.

[17] I. Held, R. Pierrehumbert, S. Garner, K. Swanson, Surface quasi-geostrophic dynamics. *J. Fluid Mech.* 282 (1995), 1–20.

[18] T. Hmidi, On the trivial solutions for the rotating patch model. *J. Evol. Eq.* 15 (2015), no. 4, 801–16.

[19] T. Hmidi, F. de la Hoz, J. Mateu, J. Verdera. Doubly connected V-states for the planar Euler equations. *SIAM J. Math. Anal.* 48 (2016), no. 3, 1892–928.

[20] T. Hmidi, J. Mateu, J. Verdera, Boundary regularity of rotating vortex patches. *Arch. Ration. Mech. Anal.* 209 (2013), no. 1, 171–208.

[21] T. Hmidi, J. Mateu, J. Verdera, On rotating doubly connected vortices. *J. Diff. Eq.* 258 (2015), no. 4, 1395–429.

[22] Z. Hassainia, T. Hmidi, F. de la Hoz, Doubly connected V-states for the generalized surface quasi-geostrophic equations. *Arch. Ration. Mech. Anal.* 220 (2016), no. 3, 1209–81.

[23] M. Juckes, Quasigeostrophic dynamics of the tropopause. *J. Armos. Sci.* (1994), 2756–68.

[24] S. Kida, Motion of an elliptical vortex in a uniform shear flow. *J. Phys. Soc. Japan* 50 (1981), 3517–20.
[25] H. Kielhöfer, *Bifurcation Theory: An Introduction with Applications to Partial Differential Equations.* Springer, New York (2011).
[26] G. Kirchhoff, *Vorlesungen uber mathematische Physik.* Teubner, Leipzig (1874).
[27] H. Lamb, *Hydrodynamics.* Dover Publications, New York (1945).
[28] G. Lapeyre, P. Klein, Dynamics of the upper oceanic layers in terms of surface quasigeostrophic theory. *J. Phys. Oceanogr.* 36 (2006), 165–76.
[29] J. Neu, The dynamics of columnar vortex in an imposed strain. *Phys. Fluids* 27 (1984), 2397–402.
[30] P. K. Newton, *The N-Vortex Problem. Analytical Techniques.* Springer, New York (2001).
[31] E. D. Rainville. *Special Functions.* Macmillan, New York (1960).
[32] J. L. Rodrigo, On the evolution of sharp fronts for the quasi-geostrophic equation. *Comm. Pure Appl. Math.* 58 (2005), no. 6, 821–66.
[33] Watson, *A Treatise on the Theory of Bessel Functions.* Cambridge University Press (1944). Trans. Amer. Math. Soc. 299 (1987), no. 2, 581–99.
[34] V. I. Yudovich, Non-stationnary flows of an ideal incompressible fluid. *Zhurnal Vych Matematika*, 3 (1963), 1032–106.

DEPARTMENT OF APPLIED MATHEMATICS AND STATISTICS AND OPERATIONS RESEARCH, FACULTY OF SCIENCE AND TECHNOLOGY, UNIVERSITY OF THE BASQUE COUNTRY UPV/EHU, BARRIO SARRIENA S/N, 48940 LEIOA, SPAIN
E-mail address: francisco.delahoz@ehu.es

IRMAR, UNIVERSITÉ DE RENNES 1, CAMPUS DE BEAULIEU, 35 042 RENNES CEDEX, FRANCE
E-mail address: zineb.hassainia@univ-rennes1.fr

IRMAR, UNIVERSITÉ DE RENNES 1, CAMPUS DE BEAULIEU, 35 042 RENNES CEDEX, FRANCE
E-mail address: thmidi@univ-rennes1.fr

5

About Least-Squares Type Approach to Address Direct and Controllability Problems

ARNAUD MÜNCH AND PABLO PEDREGAL

Abstract

We discuss the approximation of distributed null controls for partial differential equations. The main purpose is to determine an approximation of controls that drives the solution from a prescribed initial state at the initial time to the zero target at a prescribed final time. As a nontrivial example, we mainly focus on the Stokes system for which the existence of square-integrable controls have been obtained in Fursikov and Imanuvilov [*Controllability of Evolution Equations*, 1996] via Carleman-type estimates. We introduce a least-squares formulation of the controllability problem, and we show that it allows the construction of strong convergent sequences of functions toward null controls for the Stokes system. The approach consists first in introducing a class of functions satisfying *a priori* the boundary conditions in space and time – in particular the null controllability condition at time T – and then finding among this class one element satisfying the system. This second step is done by minimizing a quadratic functional, among the admissible corrector functions of the Stokes system. We also discuss briefly the direct problem for the steady Navier–Stokes system. The method does not make use of any duality arguments and therefore avoid the ill-posedness of dual methods, when parabolic type equation are considered.

1991 Mathematics Subject Classification. 65L05, 65L20, 47J30, 65D05
Key words and phrases. Controllability, approximation

Contents

Research supported in part by MTM2007-62945 of the MCyT (Spain), and
PCI08-0084-0424 of the JCCM (Castilla-La Mancha)

118

5.1 Introduction

Let $\Omega \subset \mathbb{R}^N$, $N = 2$ or $N = 3$ be a bounded connected open set whose boundary $\partial\Omega$ is Lipschitz. Let $\omega \subset \Omega$ be a (small) nonempty open subset, and assume that $T > 0$. We use the notation $Q_T = \Omega \times (0, T)$, $q_T = \omega \times (0, T)$, $\Sigma_T = \partial\Omega \times (0, T)$, and we denote by $\mathbf{n} = \mathbf{n}(\mathbf{x})$ the outward unit normal to Ω at any point $\mathbf{x} \in \partial\Omega$. Bold letters and symbols denote vector-valued functions and spaces; for instance $\mathbf{L}^2(\Omega)$ is the Hilbert space of the functions $\mathbf{v} = (v_1, \ldots, v_N)$ with $v_i \in L^2(\Omega)$ for all i.

This work is related to the null-controllability problem for the nonstationary Stokes system

$$\begin{cases} \mathbf{y}_t - \nu\Delta\mathbf{y} + \nabla\pi = \mathbf{f}\,1_\omega, & \nabla \cdot \mathbf{y} = 0 \quad \text{in } Q_T \\ \mathbf{y} = 0 \quad \text{on } \Sigma_T, \quad \mathbf{y}(\cdot, 0) = \mathbf{y}_0 \quad \text{in } \Omega \end{cases} \tag{5.1}$$

which describes a viscous incompressible fluid flow in the bounded domain Ω. We use as a control function the density of external forces $\mathbf{f} = \mathbf{f}(\mathbf{x}, t)$ concentrated in the arbitrary subdomain ω during the time interval $(0, T)$; $\mathbf{y}$ is the vector field of the fluid velocity, and π is the scalar pressure. The real ν denotes the constant viscosity of the fluid. The symbol 1_ω stands for the characteristic function of ω. We introduce the following spaces

$$\mathbf{H} = \{\boldsymbol{\varphi} \in \mathbf{L}^2(\Omega) : \nabla \cdot \boldsymbol{\varphi} = 0 \text{ in } \Omega, \ \boldsymbol{\varphi} \cdot \mathbf{n} = 0 \text{ on } \partial\Omega\},$$

$$\mathbf{V} = \{\boldsymbol{\varphi} \in \mathbf{H}_0^1(\Omega) : \nabla \cdot \boldsymbol{\varphi} = 0 \text{ in } \Omega\}, \quad U = \left\{\psi \in L^2(\Omega) : \int_\Omega \psi(\mathbf{x})\, d\mathbf{x} = 0\right\}. \tag{5.2}$$

Then, for any $\mathbf{y}_0 \in \mathbf{H}$, $T > 0$, and $\mathbf{f} \in \mathbf{L}^2(q_T)$, there exists exactly one solution $(\mathbf{y}, \pi)$ of (5.1) with the following regularity:

$$\mathbf{y} \in C^0\left([0, T]; \mathbf{H}\right) \cap L^2\left(0, T; \mathbf{V}\right), \quad \pi \in L^2(0, T; U)$$

(see [27]). The null-controllability problem for (5.1) at time T is the following:

For any $\mathbf{y}_0 \in \mathbf{H}$, find $\mathbf{f} \in \mathbf{L}^2(q_T)$ such that the corresponding solution to (5.1) satisfies

$$\mathbf{y}(\cdot, T) = \mathbf{0} \quad \text{in } \Omega. \tag{5.3}$$

The controllability properties of evolution PDEs have attracted a lot of works in the last decades: some relevant references are [3, 11, 16, 17, 28]. In particular, the Stokes – and more generally the Navier–Stokes – system has received a lot of attention: we mention [6, 15]. The following result is proved in [10] (see also [4, 11, 14]) by the way of Carleman estimates.

Theorem 5.1 (Fursikov–Imanuvilov) *The linear system (5.1) is null-controllable at any time $T > 0$.*

On the other hand, the (numerical) approximation of controls either distributed or located on the boundary for the Stokes system has received much less attention, due to the underlying ill-posedness of the approximation. In practice, such approximation is usually addressed in the framework of an optimal control reformulation. Precisely, one seeks to minimize the quadratic functional $J(\mathbf{f}) := \frac{1}{2}\|\mathbf{f}\|^2_{\mathbf{L}^2(q_T)}$ over the nonempty set

$$C(\mathbf{y}_0, T) = \{(\mathbf{y}, \mathbf{f}) : \mathbf{f} \in \mathbf{L}^2(q_T), \ \mathbf{y} \text{ solves (5.1) and satisfies (5.3)}\}.$$

Following [13], duality arguments allow to replace this constrained minimization by the unconstrained minimization of its conjugate function J^* defined as

$$J^*(\varphi_T) = \frac{1}{2} \iint_{q_T} |\varphi|^2 \, d\mathbf{x} \, dt + \int_\Omega \mathbf{y}_0 \cdot \varphi(\cdot, 0) \, d\mathbf{x}$$

over $\varphi_T \in \mathcal{H}$, where (φ, σ) solves the adjoint backward Stokes system associated with (5.1):

$$\begin{cases} -\varphi_t - \nu\Delta\varphi + \nabla\sigma = 0, & \nabla \cdot \varphi = 0 \quad \text{in } Q_T \\ \varphi = 0 \quad \text{on } \Sigma_T, & \varphi(\cdot, T) = \varphi_T \quad \text{in } \Omega. \end{cases} \tag{5.4}$$

$\mathcal{H}$ is the Hilbert space defined as the completion of any smooth space functions included in $\mathbf{H}$ for the norm $\|\varphi\|_{\mathbf{L}^2(q_T)}$. The control of minimal square-integrable norm is then given by $\mathbf{f} = \hat{\varphi} \, 1_\omega$ where $\hat{\varphi}$ is associated with the unique minimizer $\hat{\varphi}_T$ in $\mathcal{H}$ of J^* (see [3, 13]). The difficulty, when one wants to approximate such control in any finite dimensional space, that is when one likes to minimize numerically J^*, is that the space $\mathcal{H}$ is huge, in particular, contains $\cup_{s\in\mathbb{N}}\mathbf{H}^{-s}(\Omega)$, and even elements that may not be distributions. Numerical experiments do suggest that the minimizer $\hat{\varphi}_T$ is very singular (see [8, 9, 20] for a detailed analysis in the close case of the heat equation). This phenomenon is independent of the choice of J, but is related to the use of dual variables. Actually, the equality (5.3) can be viewed as an equality in a very small space (due to the strong regularization effect of the heat kernel). Accordingly, the associated multiplier φ_T belongs to a large dual space, much larger than $\mathbf{L}^2(\Omega)$, that is hard to represent (numerically) in any finite-dimensional space.

An alternative way of looking at these problems and avoiding the introduction of dual variables has been proposed in [24]. It is based on the following simple strategy. Instead of working all the time with solutions of the underlying state equation, and looking for one that may comply with the final desired state, one considers a suitable class of functions complying with required initial, boundary, final conditions, and appropriate regularity, and seeks one of those that is a solution of the state equation. This is in practice accomplished by setting up an error functional defined for all feasible functions, and measuring how far those are from being a solution of the underlying state equation.

One advantage of this variational approach is that the way to get closer to a solution of the problem is by minimizing a functional that cannot get stuck on local minima because the only critical points of the error turn out to be global minimizers with zero error. Therefore a general strategy for approximation consists in using a descent algorithm for this error functional. This approach which has the flavor of a least-squares type method has been employed successfully in our null-controllability context for the linear heat equation in [21] and for an hyperbolic system in [18].

We apply this approach to the Stokes system. In Section 5.2, we describe the ingredients of the variational approach for the system (5.1) and reduce the search of one controlled trajectory for the Stokes system to the minimization of the quadratic functional E defined by (5.23) over the affine space $\mathcal{A}$ defined by (5.5). In Section 5.3, by a general-purpose lemma (Lemma 5.9), using the very specific structure of the functional E, we prove that we may construct minimizing sequences for the error functional E that do converge strongly to an extremal point for E (see Proposition 5.8). In Section 5.4, we adapt the argument for the direct problem of the steady Navier–Stokes system.

Results of this work were partially announced in the note [22].

5.2 A Least-Squares Reformulation

Following [21, 24], we define the nonempty space

$$\mathcal{A} = \Big\{ (\mathbf{y}, \pi, \mathbf{f}); \ \mathbf{y} \in \mathbf{L}^2(0, T, \mathbf{H}_0^1(\Omega)), \mathbf{y}_t \in \mathbf{L}^2(0, T; \mathbf{H}^{-1}(\Omega)),$$

$$\mathbf{y}(\cdot, 0) = \mathbf{y}_0, \ \mathbf{y}(\cdot, T) = 0, \ \pi \in L^2(0, T; U), \ \mathbf{f} \in \mathbf{L}^2(q_T) \Big\}. \quad (5.5)$$

These hypotheses on **y** imply that it belongs to $C([0, T], \mathbf{L}^2(\Omega))$ and give a meaning to the equalities $\mathbf{y}(\cdot, 0) = \mathbf{y}_0$, $\mathbf{y}(\cdot, T) = 0$ in $\mathbf{L}^2(\Omega)$. Note also that $\mathcal{A}$ is defined in agreement with the regularity of any solution $(\mathbf{y}, \pi)$ of the Stokes system with a source term $\mathbf{f} \in \mathbf{L}^2(Q_T)$. Then, we define the functional $E: \mathcal{A} \to \mathbb{R}^+$ by

$$E(\mathbf{y}, \pi, \mathbf{f}) = \frac{1}{2} \iint_{Q_T} (|\mathbf{v}_t|^2 + |\nabla \mathbf{v}|^2 + |\nabla \cdot \mathbf{y}|^2) \, d\mathbf{x} \, dt$$

(5.6)

where the corrector **v** is the unique solution in $\mathbf{H}^1(Q_T)$ of the (elliptic) boundary value problem

$$\begin{cases} -\mathbf{v}_{tt} - \Delta \mathbf{v} + (\mathbf{y}_t - \nu \Delta \mathbf{y} + \nabla \pi - \mathbf{f} 1_\omega) = 0, & \text{in } Q_T, \\ \mathbf{v} = 0 \quad \text{on } \Sigma_T, \quad \mathbf{v}_t = 0 \quad \text{on } \Omega \times \{0, T\}. \end{cases}$$

(5.7)

For any $(\mathbf{y}, \pi, \mathbf{f}) \in \mathcal{A}$, the term $\mathbf{y}_t - \nu \Delta \mathbf{y} + \nabla \pi - \mathbf{f} 1_\omega$ belongs to $\mathbf{L}^2(0, T; \mathbf{H}^{-1}(\Omega))$ so that the functional E is well-defined in $\mathcal{A}$. The approach developed here is based on the following result.

Proposition 5.2 $(\mathbf{y}, \pi)$ is a controlled solution of the Stokes system (5.1) by the control function $\mathbf{f} 1_\omega \in \mathbf{L}^2(q_T)$ if and only if $(\mathbf{y}, \pi, \mathbf{f})$ is a solution of the extremal problem:

$$\inf_{(\mathbf{y}, \pi, \mathbf{f}) \in \mathcal{A}} E(\mathbf{y}, \pi, \mathbf{f}).$$

(5.8)

Proof From the controllability of the Stokes system given by Theorem 5.12, the extremal problem (5.25) is well-posed in the sense that the infimum, equal to zero, is reached by any controlled solution of the Stokes system. Note that, without additional assumptions, the minimizer is not unique. Conversely, we check that any minimizer for E is a solution of the (controlled) Stokes system: let $(\mathbf{Y}, \Pi, \mathbf{F}) \in \mathcal{A}_0$ be arbitrary where

$$\mathcal{A}_0 = \Big\{ (\mathbf{y}, \pi, \mathbf{f}); \ \mathbf{y} \in \mathbf{L}^2(0, T, \mathbf{H}_0^1(\Omega)), \ \mathbf{y}_t \in \mathbf{L}^2(0, T; \mathbf{H}^{-1}(\Omega)),$$

$$\mathbf{y}(\cdot, 0) = \mathbf{y}(\cdot, T) = 0, \ \pi \in L^2(0, T; U), \ \mathbf{f} \in \mathbf{L}^2(q_T) \Big\}.$$

(5.9)

The first variation of E at the point $(\mathbf{y}, \pi, \mathbf{f})$ in the admissible direction $(\mathbf{Y}, \Pi, \mathbf{F})$ defined by

$$\langle E'(\mathbf{y}, \pi, \mathbf{f}), (\mathbf{Y}, \Pi, \mathbf{F}) \rangle = \lim_{\eta \to 0} \frac{E((\mathbf{y}, \pi, \mathbf{f}) + \eta(\mathbf{Y}, \Pi, \mathbf{F})) - E(\mathbf{y}, \pi, \mathbf{f})}{\eta}, \quad (5.10)$$

exists, and is given by

$$\langle E'(\mathbf{y}, \pi, \mathbf{f}), (\mathbf{Y}, \Pi, \mathbf{F}) \rangle = \iint_{Q_T} \left(\mathbf{v}_t \cdot \mathbf{V}_t + \nabla \mathbf{v} \cdot \nabla \mathbf{V} + (\nabla \cdot \mathbf{y})(\nabla \cdot \mathbf{Y}) \right) d\mathbf{x} \, dt$$

(5.11)

where the corrector $\mathbf{V} \in \mathbf{H}^1(Q_T)$, associated with $(\mathbf{Y}, \Pi, \mathbf{F})$, is the unique solution of

$$\begin{cases} -\mathbf{V}_{tt} - \Delta \mathbf{V} + (\mathbf{Y}_t - \nu \Delta \mathbf{Y} + \nabla \Pi - \mathbf{F} 1_\omega) = 0 & \text{in } Q_T, \\ \mathbf{V} = 0 \quad \text{on } \Sigma_T, \quad \mathbf{V}_t = 0 \quad \text{on } \Omega \times \{0, T\}. \end{cases}$$

(5.12)

Multiplying the main equation of this system by $\mathbf{v}$ (recall that $\mathbf{v}$ is the corrector associated with the minimizer $(\mathbf{y}, \pi, \mathbf{f})$), integrating by parts, and using the boundary conditions on $\mathbf{v}$ and $\mathbf{V}$, we get

$$\langle E'(\mathbf{y}, \pi, \mathbf{f}), (\mathbf{Y}, \Pi, \mathbf{F}) \rangle = - \iint_{Q_T} (-\mathbf{Y} \cdot \mathbf{v}_t + \nu \nabla \mathbf{Y} \cdot \nabla \mathbf{v} - \Pi \nabla \cdot \mathbf{v} - \mathbf{F} \cdot \mathbf{v} 1_\omega) \, d\mathbf{x} \, dt$$

$$+ \iint_{Q_T} (\nabla \cdot \mathbf{y})(\nabla \cdot \mathbf{Y}) \, d\mathbf{x} \, dt, \quad \forall (\mathbf{Y}, \Pi, \mathbf{F}) \in \mathcal{A}_0,$$

(5.13)

where we have used that

$$- \int_0^T \langle \mathbf{Y}_t, \mathbf{v} \rangle_{\mathbf{H}^{-1}(\Omega), \mathbf{H}^1(\Omega)} \, dt = \iint_{Q_T} \mathbf{Y} \cdot \mathbf{v}_t \, d\mathbf{x} \, dt - \int_\Omega [\mathbf{Y} \cdot \mathbf{v}]_0^T \, d\mathbf{x}$$

$$= \iint_{Q_T} \mathbf{Y} \cdot \mathbf{v}_t \, d\mathbf{x} \, dt,$$

and that

$$\int_0^T \langle \nabla \Pi, \mathbf{v} \rangle_{\mathbf{H}^{-1}(\Omega), \mathbf{H}^1(\Omega)} \, dt = - \iint_{Q_T} \Pi \nabla \cdot \mathbf{v} \, d\mathbf{x} \, dt.$$

Therefore, if $(\mathbf{y}, \pi, \mathbf{f})$ minimizes E, the equality $\langle E'(\mathbf{y}, \pi, \mathbf{f}), (\mathbf{Y}, \Pi, \mathbf{F}) \rangle = 0$ for all $(\mathbf{Y}, \Pi, \mathbf{F}) \in \mathcal{A}_0$ implies that the corrector $\mathbf{v} = \mathbf{v}(\mathbf{y}, \pi, \mathbf{f})$ solution of (5.40) satisfies the conditions

$$\begin{cases} \mathbf{v}_t + \nu \Delta \mathbf{v} - \nabla(\nabla \cdot \mathbf{y}) = 0, \quad \nabla \cdot \mathbf{v} = 0, \quad \text{in } Q_T, \\ \mathbf{v} = 0, \quad \text{in } q_T. \end{cases}$$

(5.14)

But from the unique continuation property for the Stokes system (see [4, 5]), it turns out that $\mathbf{v} = 0$ in Q_T and that $\nabla \cdot \mathbf{y}$ is a constant in Q_T. Eventually, the relation (5.13) is then reduced to

$$\langle E'(\mathbf{y}, \pi, \mathbf{f}), (\mathbf{Y}, \Pi, \mathbf{F}) \rangle = (\nabla \cdot \mathbf{y}) \iint_{Q_T} \nabla \cdot \mathbf{Y} \, d\mathbf{x} \, dt = 0, \quad \forall (\mathbf{Y}, \Pi, \mathbf{F}) \in \mathcal{A}_0$$

124   *Arnaud Münch and Pablo Pedregal*

and then implies that this constant is zero. Consequently, if $(\mathbf{y}, \pi, \mathbf{f}) \in \mathcal{A}$ is a minimizer for E, then $\nabla \cdot \mathbf{y} = 0$ in Q_T, and the corrector $\mathbf{v}$ is zero in Q_T, so that $E(\mathbf{y}, \pi, \mathbf{f}) = 0$. Therefore, $(\mathbf{y}, \pi, \mathbf{f})$ solves (5.1), and since $(\mathbf{y}, \pi, \mathbf{f}) \in \mathcal{A}$, the state $\mathbf{y}$ is controlled at the time T by the function $\mathbf{f}$ which acts as a control distributed in ω. $\square$

Remark 5.3 The proof of Proposition 5.2 only utilizes optimality of $(\mathbf{y}, \pi, \mathbf{f})$ and not its minimality. Therefore in the statement of the proposition, we could have written instead: $(\mathbf{y}, \pi)$ is a controlled solution of the Stokes system (5.1) by the control function $\mathbf{f} 1_\omega \in \mathbf{L}^2(q_T)$ if and only if $(\mathbf{y}, \pi, \mathbf{f})$ is a stationary point for the functional $E(\mathbf{y}, \pi, \mathbf{f})$ over $(\mathbf{y}, \pi, \mathbf{f}) \in \mathcal{A}$. This is relevant from the perspective of the numerical simulation for it guarantees that the numerical procedure based on a descent strategy cannot get stuck in local minima.

Remark 5.4 For any $(\mathbf{y}, \pi, \mathbf{f}) \in \mathcal{A}$, the cost E can be formulated as follows

$$E(\mathbf{y}, \pi, \mathbf{f}) = \frac{1}{2} \|\mathbf{y}_t - \nu \Delta \mathbf{y} + \nabla \pi - \mathbf{f} 1_\omega\|^2_{\mathbf{H}^{-1}(Q_T)} + \frac{1}{2} \|\nabla \cdot \mathbf{y}\|^2_{\mathbf{L}^2(Q_T)}.$$

This justifies the least-squares terminology. The use of least-squares type approaches to solve linear and nonlinear problems is not new and we refer to [1, 2, 12] for many applications. The use of least-squares type approaches in the controllability context comes from [24].

Remark 5.5 The quasi-incompressibility case is obtained in the same way. It suffices to add $\epsilon \pi$ (for any $\epsilon > 0$) to the divergence term in the functional E. This is also in practice a classical trick to fix the constant of the pressure π (see Section 5.4).

Remark 5.6 The approach allows to consider compact support control jointly in time and space. It suffices to replace the function 1_ω in (5.1) by any compact support function in time and space, say $1_{\tilde{q}_T}$, where $\tilde{q}_T$ denotes a nonempty subset of Q_T. Since Theorem 5.12 holds for any controllability time T and any subset ω of Ω, the controllability of (5.1) remains true as soon as $\tilde{q}_T$ contains any nonempty cylindrical domain of the form $\omega_1 \times (t_1, t_2) \subset \Omega \times (0, T)$.

Remark 5.7 *A fortiori*, the approach is well-adapted to address the direct problem (which consists, $\mathbf{f}$ being fixed, in solving the boundary value problem (5.1): it suffices to remove from $\mathcal{A}$ and $\mathcal{A}_0$ the condition (5.3), and fix the forcing term $\mathbf{f}$ (see [2] using a similar least-squares type point of view). In that case, the second line of (5.14) is replaced by $\mathbf{v}(\cdot, T) = 0$, which implies with the first line, that $\mathbf{v}$ and $\nabla \cdot \mathbf{y} = 0$ both vanish in Q_T.

It is worth noting that this approach allows to treat at the same time the null-controllability constraint and the incompressibility constraint. In this sense, the pair $(\pi, \mathbf{f})$ can be regarded as a control function for the set of constraints

$$\mathbf{y}(\cdot, T) = 0 \quad \text{on } \Omega, \qquad \nabla \cdot \mathbf{y} = 0 \quad \text{in } Q_T. \tag{5.15}$$

These two conditions are compatibles: there is no competition between them. In the uncontrolled situation, from the uniqueness, the pressure π is unique as soon as the source term (here $\mathbf{f} 1_\omega$) is fixed. On the other hand, in our controllability context, the pair $(\pi, \mathbf{f})$ is not unique: the pressure π depends on $\mathbf{f} 1_\omega$ and vice versa. Therefore, the optimization with respect to both variables at the same time makes sense. From this point of view, we may reformulate the problem as a general controllability problem for the heat equation:

$$\mathbf{y}_t - \nu \Delta \mathbf{y} = V := \mathbf{f} 1_\omega - \nabla \pi \quad \text{in } Q_T,$$

V being a control function such that (5.15) holds. The control function V acts on the whole domain, but on the other hand, should take the specific form $V := \mathbf{f} 1_\omega - \nabla \pi$.

Again, this perspective is different with the classical one, which consists of finding a control $\mathbf{v} \in \mathbf{L}^2(q_T)$, such that $\mathbf{y}(\cdot, T) = 0$ in Ω where $(\mathbf{y}, \pi)$ solves (5.1). This can done by duality, penalization technique, etc. Conversely, one may also consider iteratively first the null-controllability constraint, that is, for any π fixed in $L^2(0, T; U)$, find a control $\mathbf{f}_\pi 1_\omega$ such that (5.3) holds, and then find the pressure π such that $\nabla \cdot \mathbf{y}_\pi = 0$ holds in Q_T. Using again a least-squares type approach (for the heat equation, as developed in [21]), the first step reduces to solve, for any fixed $\pi \in L^2(0, T, U)$, the problem

$$\inf_{(\mathbf{y}_\pi, \mathbf{f}_\pi) \in \mathcal{A}_1} \tilde{E}(\mathbf{y}, \mathbf{f}) := \frac{1}{2} \|\mathbf{v}\|^2_{\mathbf{H}^1(Q_T)}$$

where $\mathbf{v} = \mathbf{v}(\mathbf{y}_\pi, \pi, \mathbf{f}_\pi)$ solves (5.40) and $\mathcal{A}_1$ is given by

$$\mathcal{A}_1 = \Big\{ (\mathbf{y}, \mathbf{f}); \ \mathbf{y} \in \mathbf{L}^2(0, T, \mathbf{H}_0^1(\Omega)), \mathbf{y}_t \in \mathbf{L}^2(0, T; \mathbf{H}^{-1}(\Omega)),$$

$$\mathbf{y}(\cdot, 0) = \mathbf{y}_0, \ \mathbf{y}(\cdot, T) = 0, \ \mathbf{f} \in \mathbf{L}^2(q_T) \Big\}.$$

The second step consists in updating the pressure according to a descent direction for the function $G : L^2(0, T, U) \to \mathbb{R}$ defined by $G(\pi) := 1/2 \|\nabla \cdot \mathbf{y}_\pi\|^2_{L^2(Q_T)}$. We get that the first variation of G at π in the direction

$\overline{\pi} \in L^2(0, T; U)$ is given by $<G'(\pi), \overline{\pi}> = \iint_{Q_T} \nabla \overline{\pi} \cdot \mathbf{p} \, dx \, dt$ where $\mathbf{p}$ solves

$$-\mathbf{p}_t - \nu \Delta \mathbf{p} = \nabla(\nabla \cdot \mathbf{y}_\pi) \quad \text{in } Q_T, \quad \mathbf{p}(\cdot, T) = 0 \quad \text{in } \Omega, \quad \mathbf{p} = 0 \quad \text{on } \Sigma_T.$$

Again, this direct problem may be solved within the variational approach developed in this work (see Remark 5.7).

5.3 Convergence of Some Minimizing Sequences for E

Proposition 5.2 reduces the approximation of a null control for (5.1) to a minimization of the functional E over $\mathcal{A}$. As a preliminary step, since $\mathcal{A}$ is not a vectorial space, we remark that any element of $\mathcal{A}$ can be written as the sum of one element of $\mathcal{A}$, say $s_{\mathcal{A}}$, plus any element of the vectorial space $\mathcal{A}_0$. Thus, we consider for any $s_{\mathcal{A}} := (\mathbf{y}_{\mathcal{A}}, \pi_{\mathcal{A}}, \mathbf{f}_{\mathcal{A}}) \in \mathcal{A}$ the following problem:

$$\min_{(\mathbf{y}, \pi, \mathbf{f}) \in \mathcal{A}_0} E_{s_{\mathcal{A}}}(\mathbf{y}, \pi, \mathbf{f}), \qquad E_{s_{\mathcal{A}}}(\mathbf{y}, \pi, \mathbf{f}) := E(s_{\mathcal{A}} + (\mathbf{y}, \pi, \mathbf{f})). \qquad (5.16)$$

Problems (5.16) and (5.25) are equivalent. Any solution of Problem (5.16) is a solution of the initial problem (5.25). Conversely, any solution of Problem (5.25) can be decomposed as the sum $s_{\mathcal{A}} + s_{\mathcal{A}_0}$, for some $s_{\mathcal{A}_0}$ in $\mathcal{A}_0$.

We endow the vectorial space $\mathcal{A}_0$ with its natural norm $\| \cdot \|_{\mathcal{A}_0}$ such that:

$$\|\mathbf{y}, \pi, \mathbf{f}\|_{\mathcal{A}_0}^2 := \iint_{Q_T} (|\mathbf{y}|^2 + |\nabla \mathbf{y}|^2) \, dx \, dt + \int_0^T \|\mathbf{y}_t(\cdot, t)\|_{\mathbf{H}^{-1}(\Omega)}^2 \, dt$$

$$+ \iint_{Q_T} (|\mathbf{f}|^2 + |\pi|^2) \, dx \, dt, \qquad (5.17)$$

recalling that $\|\mathbf{y}_t\|_{\mathbf{H}^{-1}(\Omega)} = \|\mathbf{g}\|_{\mathbf{H}_0^1}(\Omega)$ where $\mathbf{g} \in \mathbf{H}_0^1(\Omega)$ solves $-\Delta \mathbf{g} = \mathbf{y}_t$ in Ω. We denote $\langle , \rangle_{\mathcal{A}_0}$ the corresponding scalar product. $(\mathcal{A}_0, \| \cdot \|_{\mathcal{A}_0})$ is an Hilbert space.

The relation (5.13) allows to define a minimizing sequence in $\mathcal{A}_0$ for $E_{s_{\mathcal{A}}}$.

It turns out that minimizing sequences for $E_{s_{\mathcal{A}}}$ which belong to a precise subset of $\mathcal{A}_0$ remain bounded uniformly. This very valuable property is not *a priori* guaranteed from the definition of $E_{s_{\mathcal{A}}}$. The boundedness of $E_{s_{\mathcal{A}}}$ implies only the boundedness of the corrector $\mathbf{v}$ for the $\mathbf{H}^1(Q_T)$-norm and the boundedness of the divergence $\nabla \cdot \mathbf{y}$ of the velocity field for the $L^2(Q_T)$ norm. Actually, this property is due to the fact that the functional $E_{s_{\mathcal{A}}}$ is invariant in the subset of $\mathcal{A}_0$ which satisfies the state equations of (5.1).

In order to construct a minimizing sequence bounded in $\mathcal{A}_0$ for E_{s_A}, we introduce the linear continuous operator $\mathbf{T}$ which maps a triplet $(\mathbf{y}, \pi, \mathbf{f}) \subset \mathcal{A}$ into the corresponding vector $(\mathbf{v}, \nabla \cdot \mathbf{y}) \in \mathbf{H}^1(Q_T) \times L^2(Q_T)$, with the corrector $\mathbf{v}$ as defined by (5.40). Then we define the space $A = \mathrm{Ker}\,\mathbf{T} \cap \mathcal{A}_0$ composed of the elements $(\mathbf{y}, \pi, \mathbf{f})$ satisfying the Stokes system and such that $\mathbf{y}$ vanishes on the boundary ∂Q_T. Note that A is not the trivial space: it suffices to consider the difference of two distinct null-controlled solutions of (5.1). Finally, we note $A^\perp = (\mathrm{Ker}\,\mathbf{T} \cap \mathcal{A}_0)^\perp$ the orthogonal complement of A in $\mathcal{A}_0$ and $P_{A^\perp} : \mathcal{A}_0 \to A^\perp$ the (orthogonal) projection on $A^\perp$.

We then define the following minimizing sequence $(\mathbf{y}^k, \pi^k, \mathbf{f}^k)_{k \geq 0} \in A^\perp$ as follows:

$$\begin{cases} (\mathbf{y}^0, \pi^0, \mathbf{f}^0) \text{ given in } A^\perp, \\ (\mathbf{y}^{k+1}, \pi^{k+1}, \mathbf{f}^{k+1}) = (\mathbf{y}^k, \pi^k, \mathbf{f}^k) - \eta_k P_{A^\perp}(\overline{\mathbf{y}}^k, \overline{\pi}^k, \overline{\mathbf{f}}^k), \quad k \geq 0 \end{cases} \tag{5.18}$$

where $(\overline{\mathbf{y}}^k, \overline{\pi}^k, \overline{\mathbf{f}}^k) \in \mathcal{A}_0$ is defined as the unique solution of the formulation

$$\langle (\overline{\mathbf{y}}^k, \overline{\pi}^k, \overline{\mathbf{f}}^k), (\mathbf{Y}, \Pi, \mathbf{F}) \rangle_{\mathcal{A}_0} = \langle E'_{s_0}(\mathbf{y}^k, \pi^k, \mathbf{f}^k), (\mathbf{Y}, \Pi, \mathbf{F}) \rangle, \quad \forall (\mathbf{Y}, \Pi, \mathbf{F}) \in \mathcal{A}_0. \tag{5.19}$$

η_k denotes a positive descent step. In particular, (5.19) implies that $\overline{\pi}^k = -\nabla \cdot \mathbf{v}^k \in L^2(Q_T)$ and $\overline{\mathbf{f}}^k = -\mathbf{v}^k \mathbf{1}_\omega \in L^2(q_T)$ (actually in $\mathbf{H}^1(q_T)$).

One main issue of our variational approach is to establish the convergence of the minimizing sequence defined by (5.18). We have the following result.

Proposition 5.8 For any $s_A \in \mathcal{A}$ and any $\{\mathbf{y}^0, \pi^0, \mathbf{f}^0\} \in A^\perp$, the sequence $s_A + \{(\mathbf{y}^k, \pi^k, \mathbf{f}^k)\}_{k \geq 0} \in \mathcal{A}$ converges strongly to a solution of the extremal problem (5.25).

This proposition is the consequence of the following abstract result which can be adapted to many different situations where this variational perspective can be of help.

Lemma 5.9 Suppose $\mathbf{T} : X \mapsto Y$ is a linear, continuous operator between Hilbert spaces, and $H \subset X$, a closed subspace, $u_0 \in X$. Put

$$E : u_0 + H \mapsto \mathbb{R}^+, \quad E(u) = \frac{1}{2}\|\mathbf{T}u\|^2, \quad A = \mathrm{Ker}\,\mathbf{T} \cap H.$$

1. $E : u_0 + A^\perp \to \mathbb{R}$ is quadratic, nonnegative, and strictly convex, where $A^\perp$ is the orthogonal complement of A in H.

2. If we regard E as a functional defined on H, $E(u_0 + \cdot)$, and identify H with its dual, then the derivative $E'(u_0 + \cdot)$ always belongs to $A^\perp$. In particular, a typical steepest descent procedure for $E(u_0 + \cdot)$ will always stay in the manifold $u_0 + A^\perp$.
3. If, in addition, $\min_{u \in H} E(u_0 + u) = 0$, then the steepest descent scheme will always produce sequences converging (strongly in X) to a unique (in $u_0 + A^\perp$) minimizer $u_0 + \bar{u}$ with zero error.

Proof of Lemma 5.9 Suppose there are $u_i \in A^\perp$, $i = 1, 2$, such that

$$E\left(u_0 + \frac{1}{2}u_1 + \frac{1}{2}u_2\right) = \frac{1}{2}E(u_0 + u_1) + \frac{1}{2}E(u_0 + u_2).$$

Due to the strict convexity of the norm in a Hilbert space, we deduce that this equality can only occur if $\mathbf{T}u_1 = \mathbf{T}u_2$. So therefore $u_1 - u_2 \in A \cap A^\perp = \{0\}$, and $u_1 = u_2$. For the second part, note that for arbitrary $U \in A$, $\mathbf{T}U = 0$, and so

$$E(u_0 + u + U) = \frac{1}{2}\|\mathbf{T}u_0 + \mathbf{T}u + \mathbf{T}U\|^2 = \frac{1}{2}\|\mathbf{T}u_0 + \mathbf{T}u\|^2 = E(u_0 + u).$$

Therefore the derivative $E'(u_0 + u)$, the steepest descent direction for E at $u_0 + u$, has to be orthogonal to all such $U \in A$.

Finally, assume $E(u_0 + \bar{u}) = 0$. It is clear that this minimizer is unique in $u_0 + A^\perp$ (recall the strict convexity in (i)). This, in particular, implies that for arbitrary $u \in A^\perp$,

$$\langle E'(u_0 + u), \bar{u} - u \rangle \leq 0, \tag{5.20}$$

because this inner product is the derivative of the section $t \mapsto E(u_0 + t\bar{u} + (1-t)u)$ at $t = 0$, and this section must be a positive parabola with the minimum point at $t = 1$. If we consider the gradient flow

$$u'(t) = -E'(u_0 + u(t)), \quad t \in [0, +\infty),$$

then, because of (5.20),

$$\frac{d}{dt}\left(\frac{1}{2}\|u(t) - \bar{u}\|^2\right) = \langle u(t) - \bar{u}, u'(t) \rangle = \langle u(t) - \bar{u}, -E'(u_0 + u(t)) \rangle \leq 0.$$

This implies that sequences produced through a steepest descent method will be minimizing for E, uniformly bounded in X (because $\|u(t) - \bar{u}\|$ is a nonincreasing function of t), and due to the strict convexity of E restricted to $u_0 + A^\perp$, they will have to converge towards the unique minimizer $u_0 + \bar{u}$. $\qquad\square$

Remark 5.10 Despite the strong convergence in this statement, it may not be true that the error is coercive, even restricted to $u_0 + A^{\perp}$, so that strong convergence could be very slow. Because of this same reason, it may be impossible to establish rates of convergence for these minimizing sequences.

The element u_0 determines the nonhomogeneous data set of each problem: source term, boundary conditions, initial and/or final condition, etc. The subspace H is the subset of the ambient Hilbert space X for which the data set vanishes. $\mathbf{T}$ is the operator defining the corrector, so that $\operatorname{Ker}\mathbf{T}$ is the subspace of all solutions of the underlying equation or system. The subspace A is the subspace of all solutions of the problem with vanishing data set. In some situations A will be trivial, but in some others will not be so. The important property is (iii) in the statement guaranteeing that we indeed have strong convergence in X of iterates. The main requirement for this to hold is to know, *a priori*, that the error attains its minimum value zero somewhere, which in the situation treated here is guaranteed by Theorem 5.12.

Proof of Proposition 5.8 The result is obtained by applying Lemma 5.9 as follows. If we put $B = \{\mathbf{y} \in L^2(0, T, \mathbf{H}_0^1(\Omega)) : \mathbf{y}_t \in L^2(0, T; \mathbf{H}^{-1}(\Omega))\}$, X is taken to be $B \times L^2(0, T; U) \times L^2(q_T)$. H is taken to be $\mathcal{A}_0$ as given in (5.9) and $u_0 = s_{\mathcal{A}} \in \mathcal{A} \subset X$. The operator $\mathbf{T}$ maps a triplet $(\mathbf{y}, \pi, \mathbf{f}) \in \mathcal{A} \subset X$ into $(\mathbf{v}, \nabla \cdot \mathbf{y}) \in Y := \mathbf{H}^1(Q_T) \times L^2(Q_T)$ as explained earlier.

Remark 5.11 The construction of the minimizing sequence only requires the resolution of standard well-posed elliptic problems over Q_T, well-adapted to general situations (time-dependent support, mesh adaptation, etc.). On the other hand, it is important to highlight that the $L^2(q_T)$ control function $\mathbf{f}$ obtained from the minimizing procedure does not *a priori* minimize any specific norm (for instance the L^2-norm).

Without the projection on $(\operatorname{Ker}\mathbf{T} \cap \mathcal{A}_0)^{\perp}$ in (5.18), the sequence $(\mathbf{y}^k, \pi^k, \mathbf{f}^k)$ remains a minimizing sequence for E_{s_A}: actually, the values of the cost E_{s_A} along the sequence $(\mathbf{y}^k, \pi^k, \mathbf{f}^k)$ are equal with or without the projection. This is due to the fact that the component of the descent direction $(\overline{\mathbf{y}}^k, \overline{\pi}^k, \overline{\mathbf{f}}^k)$ on $(\operatorname{Ker}\mathbf{T} \cap \mathcal{A}_0)$ does not affect the value of the cost: on the other hand, without the projection, the minimizing sequence may not be bounded uniformly in $\mathcal{A}_0$, in particular the control function $\mathbf{f}$ may not be bounded in $L^2(q_T)$.

The subset $A^\perp$ is not explicit, so that in practice the projection $P_{A^\perp}(\overline{\mathbf{y}}^k, \overline{\pi}^k, \overline{\mathbf{f}}^k)$ may be defined by $P_{A^\perp}(\overline{\mathbf{y}}^k, \overline{\pi}^k, \overline{\mathbf{f}}^k) = (\overline{\mathbf{y}}^k, \overline{\pi}^k, \overline{\mathbf{f}}^k) - \boldsymbol{p}$, where $\boldsymbol{p}$ solves the extremal problem:

$$\min_{\boldsymbol{p} \in A} \|\boldsymbol{p} - (\overline{\mathbf{y}}^k, \overline{\pi}^k, \overline{\mathbf{f}}^k)\|_{A_0}. \tag{5.21}$$

Recalling that A is by definition the set of triplets $(\mathbf{y}, \pi, \mathbf{f})$ satisfying $\mathbf{y}_t - \nu \Delta \mathbf{y} + \nabla \pi - \mathbf{f} 1_\omega = 0$, $\nabla \cdot \mathbf{y} = 0$ in Q_T such that $\mathbf{y}$ vanishes on ∂Q_T, this extremal problem is nothing else than a controllability problem for the Stokes system, similar to the one considered in this work. Therefore, we shall bypass this projection and shall introduce instead a stopping criteria for the descent method measuring how far from $A^\perp$ the descent direction is.

We refer to [19] for numerical experiments.

5.4 Direct Problem for the Steady Navier–Stokes System

The least-squares approach allows to address nonlinear problem. We consider here simply the steady Navier–Stokes and address the direct problem: find $(\mathbf{y}, \pi)$ solution of

$$\begin{cases} -\nu \Delta \mathbf{y} + (\mathbf{y} \cdot \nabla)\mathbf{y} + \nabla \pi = \mathbf{f}, & \nabla \cdot \mathbf{y} = 0 \quad \text{in } \Omega \\ \mathbf{y} = 0 \quad \text{on } \partial \Omega. \end{cases} \tag{5.22}$$

We recall the following result (see [27]):

Theorem 5.12 *For any* $\mathbf{f} \in \mathbf{H}^{-1}(\Omega)$*, there exists at least one* $(\mathbf{y}, \pi) \in \mathbf{H}_0^1(\Omega) \times L_0^2(\Omega)$ *solution of (5.22). Moreover, if* $\nu^{-2}\|\mathbf{f}\|_{\mathbf{H}^{-1}(\Omega)}$ *is small enough, then the couple* $(\mathbf{y}, \pi)$ *is unique.*

In order to solve this boundary value problem, we use a least-squares type approach. We consider the space $\mathcal{A} = \mathbf{H}_0^1(\Omega) \times L^2(\Omega)$ and then we define the functional $E : \mathcal{A} \to \mathbb{R}^+$ by

$$E(\mathbf{y}, \pi) := \frac{1}{2} \int_\Omega (|\nabla \mathbf{v}|^2 + |\nabla \cdot \mathbf{y}|^2) \, dx \tag{5.23}$$

where the corrector $\mathbf{v}$ is the unique solution in $\mathbf{H}_0^1(Q_T)$ of the (elliptic) boundary value problem

$$\begin{cases} -\Delta \mathbf{v} + (-\nu \Delta \mathbf{y} + div(\mathbf{y} \otimes \mathbf{y}) + \nabla \pi - \mathbf{f}) = 0, & \text{in } \Omega, \\ \mathbf{v} = 0 \quad \text{on } \partial \Omega. \end{cases} \tag{5.24}$$

We then consider the following extremal problem

$$\inf_{(\mathbf{y},\pi)\in\mathcal{A}} E(\mathbf{y},\pi). \tag{5.25}$$

We recall the following equality.

Lemma 5.13
- For all $\mathbf{y},\mathbf{z}$, $div(\mathbf{y}\otimes\mathbf{z})=\mathbf{y}\nabla\cdot\mathbf{z}+(\nabla\mathbf{y})\mathbf{z}$.
- For all $\mathbf{y},\mathbf{z},\mathbf{p}\in\mathbf{H}_0^1(\Omega)$,

$$\int_\Omega\left(div(\mathbf{y}\otimes\mathbf{z})+div(\mathbf{z}\otimes\mathbf{y})\right)\cdot\mathbf{p}=-\int_\Omega(\mathbf{y}\otimes\mathbf{z}+\mathbf{z}\otimes\mathbf{y}):\nabla\mathbf{p}$$

$$=-\int_\Omega\left((\nabla\mathbf{p}+(\nabla\mathbf{p})^T)\mathbf{y}\right)\cdot\mathbf{z}. \tag{5.26}$$

From Theorem 5.12, the infimum is equal to zero and is reached by an element solution of (5.22). Conversely, we would like to state that the only critical point for E corresponds to solution of (5.22).

In view of [26, corollary 1.8], let us first prove the following result.

Proposition 5.14 *E is an error functional, that is $E:\mathcal{A}\to\mathbb{R}$ is a C^1-functional over the Hilbert space $\mathcal{A}$, E is nonnegative and*

$$\lim E(\mathbf{y},\pi)=0 \quad as \quad E'(\mathbf{y},\pi)\to 0.$$

Proof We write that

$$E'(\mathbf{y},\pi)\cdot(\mathbf{Y},\Pi)=\int_\Omega\nabla\mathbf{v}\cdot\nabla\mathbf{V}+(\nabla\cdot\mathbf{y})(\nabla\cdot\mathbf{Y})\,dx \tag{5.27}$$

where $\mathbf{V}\in\mathbf{H}_0^1(\Omega)$ solves

$$\begin{cases}-\Delta\mathbf{V}+(-\nu\Delta\mathbf{Y}+div(\mathbf{y}\otimes\mathbf{Y})+div(\mathbf{Y}\otimes\mathbf{y})+\nabla\Pi)=0, & in\ \Omega, \\ \mathbf{V}=0 \quad on\ \partial\Omega.\end{cases} \tag{5.28}$$

Multiplying the state equation by $\mathbf{v}$ and using (5.13), we get

$$E'(\mathbf{y},\pi)\cdot(\mathbf{Y},\Pi)=\int_\Omega\nu\Delta\mathbf{v}\cdot\mathbf{Y}+(\nabla\mathbf{v}+(\nabla\mathbf{v})^T)\mathbf{y})\cdot\mathbf{Y}+(\nabla\cdot\mathbf{v})\Pi dx$$

$$+\int_\Omega(\nabla\cdot\mathbf{y})(\nabla\cdot\mathbf{Y})dx \tag{5.29}$$

or equivalently

$$E'(\mathbf{y}, \pi) \cdot (\mathbf{Y}, \Pi) = \int_\Omega -\nu \nabla \mathbf{v} \cdot \nabla \mathbf{Y} + (\mathbf{y} \otimes \mathbf{Y} + \mathbf{Y} \otimes \mathbf{y}) : \nabla \mathbf{v} + (\nabla \cdot \mathbf{v})\Pi \, d\mathbf{x}$$

$$+ \int_\Omega (\nabla \cdot \mathbf{y})(\nabla \cdot \mathbf{Y}) \, d\mathbf{x}. \tag{5.30}$$

We then check that we can take $\mathbf{Y} = \mathbf{v}$, i.e., $\mathbf{v} = \mathbf{v}(\mathbf{y}, \pi) \in \mathbf{H}_0^1(\Omega)$ uniquely given by the solution of (5.40) remains bounded with respect to $(\mathbf{y}, \pi) \in \mathbf{H}_0^1 \times L^2(\Omega)$. By definition, the corrector $\mathbf{v}$ solves the variational formulation:

$$\int_\Omega ((\nabla \mathbf{v} + \nu \nabla \mathbf{y} - \mathbf{y} \otimes \mathbf{y}) : \nabla \mathbf{w} - \pi \nabla \cdot \mathbf{w} - \mathbf{f} \cdot \mathbf{w}) = 0, \quad \forall \mathbf{w} \in \mathbf{H}_0^1(\Omega). \tag{5.31}$$

Taking $\mathbf{w} = \mathbf{v}$, we get that

$$\int_\Omega |\nabla \mathbf{v}|^2 \, d\mathbf{x} = \int_\Omega (\mathbf{y} \otimes \mathbf{y} : \nabla \mathbf{v} - \nu \nabla \mathbf{y} : \nabla \mathbf{v} + \pi \nabla \cdot \mathbf{v} + \mathbf{f} \cdot \mathbf{v}) \, d\mathbf{x} \tag{5.32}$$

so that, in view of the Poincaré inequality,

$$\|\mathbf{v}\|_{\mathbf{H}_0^1(\Omega)} \leq C(\|\mathbf{y} \otimes \mathbf{y}\|_{L^2(\Omega)} + \|\mathbf{y}\|_{\mathbf{H}_0^1(\Omega)} + \|\pi\|_{L^2(\Omega)} + \|\mathbf{f}\|_{\mathbf{H}^{-1}(\Omega)}), \tag{5.33}$$

leading to the result. Then, taking $\mathbf{Y} = \mathbf{v}$ in (5.30), we get

$$E'(\mathbf{y}, \pi) \cdot (\mathbf{v}, \Pi) = \int_\Omega -\nu |\nabla \mathbf{v}|^2 - (\mathbf{y} \otimes \mathbf{v} + \mathbf{v} \otimes \mathbf{y}) : \nabla \mathbf{v} + (\nabla \cdot \mathbf{v})\Pi \, d\mathbf{x}$$

$$+ \int_\Omega (\nabla \cdot \mathbf{y})(\nabla \cdot \mathbf{v}) \, d\mathbf{x}$$

$$= \int_\Omega -\nu |\nabla \mathbf{v}|^2 - (\mathbf{v} \otimes \mathbf{v}) : \nabla \mathbf{y} + \frac{1}{2}(\nabla \cdot \mathbf{y})|\mathbf{v}|^2 \, d\mathbf{x}$$

$$+ \int_\Omega (\nabla \cdot \mathbf{v})(\nabla \cdot \mathbf{y} + \mathbf{y} \cdot \mathbf{v} + \Pi) \, d\mathbf{x}. \tag{5.34}$$

Similarly, in view of (5.33), $\Pi_s = -(\nabla \cdot \mathbf{y} + \mathbf{y} \cdot \mathbf{v}) \in L^2(\Omega)$ remains bounded with respect to $(\mathbf{y}, \pi)$ and we write

$$E'(\mathbf{y}, \pi) \cdot (\mathbf{v}, \Pi_s) = \int_\Omega -\nu |\nabla \mathbf{v}|^2 - (\mathbf{v} \otimes \mathbf{v}) : \nabla \mathbf{y} + \frac{1}{2}(\nabla \cdot \mathbf{y})|\mathbf{v}|^2 \, d\mathbf{x}. \tag{5.35}$$

We then use the following result (consequence of the well-posedness of the Oseen equation):

Lemma 5.15 For any $\mathbf{y} \in \mathbf{H}_0^1(\Omega)$, $\mathbf{F} \in L^2(\Omega)$, there exists $(\mathbf{Y}, \Pi) \in H_0^1(\Omega) \times L^2(\Omega)$ with $\nabla \cdot \mathbf{Y} = 0$ such that

$$\int_\Omega (\nu \nabla \mathbf{Y} - (\mathbf{Y} \otimes \mathbf{y} + \mathbf{y} \otimes \mathbf{Y})) : \nabla \mathbf{w} - \Pi \nabla \cdot \mathbf{w} - \mathbf{F} \cdot \mathbf{w} = 0, \quad \forall \mathbf{w} \in H_0^1(\Omega)$$

(5.36)

such that $\|\mathbf{Y}, \Pi\|_{\mathbf{H}_0^1(\Omega) \times L^2(\Omega)} \le C(\|\mathbf{y}\|_{\mathbf{H}_0^1(\Omega)} + \|\mathbf{F}\|_{L^2(\Omega)})$ for some $C > 0$.

Using this lemma for $\mathbf{F} = \mathbf{v}$ and $\mathbf{w} = \mathbf{v}$ ($\mathbf{v}$ is the corrector associated with the pair $(\mathbf{y}, \pi)$), we obtain that $(\mathbf{Y}, \Pi) \in \mathbf{H}_0^1(\Omega) \times L^2(\Omega)$ satisfies $\nabla \cdot \mathbf{Y} = 0$ and

$$\int_\Omega (\nu \nabla \mathbf{Y} - (\mathbf{Y} \otimes \mathbf{y} + \mathbf{y} \otimes \mathbf{Y})) : \nabla \mathbf{v} - \Pi \nabla \cdot \mathbf{v} - \mathbf{v} \cdot \mathbf{v} = 0, \quad \forall \mathbf{w} \in H_0^1(\Omega).$$

(5.37)

With this pair $(\mathbf{Y}, \Pi)$ bounded with respect to $\mathbf{v}$ and to $\mathbf{y}$, and so with respect to $(\mathbf{y}, \pi)$, we have from (5.30), (reminder that $\nabla \cdot \mathbf{Y} = 0$)

$$E'(\mathbf{y}, \pi) \cdot (\mathbf{Y}, \Pi) = \int_\Omega -\nu \nabla \mathbf{v} \cdot \nabla \mathbf{Y} + (\mathbf{y} \otimes \mathbf{Y} + \mathbf{Y} \otimes \mathbf{y}) : \nabla \mathbf{v} + (\nabla \cdot \mathbf{v})\Pi \, d\mathbf{x}.$$

(5.38)

The property $E'(\mathbf{y}, \pi) \cdot (\mathbf{Y}, \Pi) \to 0$ then implies that $\|\mathbf{v}\|_{L^2(\Omega)} \to 0$. Then, from (5.35), the property $E'(\mathbf{y}, \pi) \cdot (\mathbf{v}, \Pi_s) \to 0$ then implies from the equality (5.37) that $\|\nabla \mathbf{v}\|_{L^2(\Omega)} \to 0$.

Then, 5.30 implies that $\int_\Omega \nabla \cdot \mathbf{y} \nabla \cdot \mathbf{Y} \, d\mathbf{x} \to 0$ for all $\mathbf{Y} \in H_0^1(\Omega)$ so that $\|\nabla \cdot \mathbf{y}\|_{L^2(\Omega)} \to 0$. $\qquad\square$

We then recall the following result:

Lemma 5.16 [26, corollary 1.8] Let H be a Hilbert space. Suppose $J : H \to \mathbb{R}$ is an error functional for which there is a unique $u^0 \in H$ with $J(u^0) = 0$. If $J(u^j) \to 0$, then $u^j \to u^0$ strongly in H, and J complies with the Palais–Smale condition.

In view of Proposition 5.14 and of the previous Lemma, we have the following result.

Proposition 5.17 If $\nu^{-1}\|\mathbf{f}\|_{H^{-1}(\Omega)}$ is small enough, then any minimizing sequence $(\mathbf{y}_k, \pi_k)_{k>0}$ for E converges strongly toward $(\mathbf{y}, \pi)$, unique solution of the boundary value problem (5.22).

Proof It suffices to apply the previous for $J = E$ and $H = \mathcal{A}$. The uniqueness here follows from Theorem 5.12.   □

Remark 5.18 We write again that

$$E'(\mathbf{y}, \pi) \cdot (\mathbf{Y}, \Pi) = \int_\Omega \nabla \mathbf{v} \cdot \nabla \mathbf{V} + (\nabla \cdot \mathbf{y})(\nabla \cdot \mathbf{Y}) \, dx \qquad (5.39)$$

where $\mathbf{V} \in \mathbf{H}_0^1(\Omega)$ solves the well-posed formulation

$$\begin{cases} -\Delta \mathbf{V} + (-\nu \Delta \mathbf{Y} + div(\mathbf{y} \otimes \mathbf{Y}) + div(\mathbf{Y} \otimes \mathbf{y}) + \nabla \Pi) = 0, & \text{in } \Omega, \\ \mathbf{V} = 0 \quad \text{on } \partial\Omega. \end{cases}$$
$$(5.40)$$

Multiplying the state equation by $\mathbf{v}$ and using (5.13), we get

$$E'(\mathbf{y}, \pi) \cdot (\mathbf{Y}, \Pi) = \int_\Omega \nu \Delta \mathbf{v} \cdot \mathbf{Y} + (\nabla \mathbf{v} + (\nabla \mathbf{v})^T)\mathbf{y}) \cdot \mathbf{Y} + (\nabla \cdot \mathbf{v})\Pi \, dx$$
$$+ \int_\Omega (\nabla \cdot \mathbf{y})(\nabla \cdot \mathbf{Y}) \, dx \qquad (5.41)$$

so that, at the optimality, the corrector function $\mathbf{v}$ solves the following linear boundary problem:

$$\begin{cases} \nu \Delta \mathbf{v} + (\nabla \mathbf{v} + (\nabla \mathbf{v})^T)\mathbf{y} - \nabla(\nabla \cdot \mathbf{y}) = 0, & \nabla \cdot \mathbf{v} = 0 \quad \text{in } \Omega \\ \mathbf{v} = 0 \quad \text{on } \partial\Omega. \end{cases}$$
$$(5.42)$$

Proposition 5.14 implies that if $E'(\mathbf{y}, \pi) = 0$, then $E(\mathbf{y}, \pi) = 0$. Therefore, the formulation (5.42) implies that $\mathbf{v} = 0$ and $\nabla \cdot \mathbf{y} = 0$. This may be seen as a unique continuation property for Navier–Stokes equation.

Remark 5.19 The previous results hold true if we replace the cost E by

$$E_\varepsilon(\mathbf{y}, \pi) := \frac{1}{2} \int_\Omega (|\nabla \mathbf{v}|^2 + |\nabla \cdot \mathbf{y} + \varepsilon \pi|^2) \, dx \qquad (5.43)$$

for any $\varepsilon > 0$, where $(\mathbf{y}, \pi)$ solves (5.40).

As a very interesting consequence, Proposition 5.17 reduces the approximation of the vanishing elements of E to the construction of one arbitrary minimizing sequence, say $(\mathbf{y}_k, \pi_k)_{k>0}$, for E.

The unsteady case as well as the controllability case are more involved: we refer to [23] based on [7, 14, 25].

References

[1] P.B. Bochev and M.D. Gunzburger, Analysis of least squares finite element methods for the Stokes equations, *Math. Comput.*, 63(208) (1994), 479–506.

[2] B. Bochev and M. Gunzburger, *Least-Squares Finite Element Methods*, Applied Mathematical Sciences, 166. Springer, New York, 2009, xxii+660 pp.

[3] J.M. Coron, *Control and Nonlinearity*, Mathematical Surveys and Monographs, AMS, Vol. 136, 2007.

[4] C. Fabre, Uniqueness results for Stokes equations and their consequences in linear and nonlinear control problems. *ESAIM:COCV*, 1 (1995/6), 267–302.

[5] C. Fabre and G. Lebeau, Prolongement unique des solutions de l'équation de Stokes (French), *Comm. PDE*, 21 (1996), 573–96.

[6] E. Fernández-Cara, S. Guerrero, O.Yu. Imanuvilov, and J.-P. Puel, Local exact controllability of the Navier–Stokes system, *J. Math. Pures Appl.*, 83(12)(2004), 1501–42.

[7] E. Fernández-Cara and A. Münch, Numerical null controllability of semilinear 1D heat equations: Fixed points, least squares and Newton methods, *Math. Control Relat. Fields*, 2(3) (2012), 217–46.

[8] E. Fernández-Cara and A. Münch, Numerical null controllability of the 1D heat equation: Primal algorithms, *Séma J.*, 61(1) (2013), 49–78.

[9] E. Fernández-Cara and A. Münch, Numerical null controllability of the 1D heat equation: Duality and Carleman weights. *J. Optim. Theory Appl.*, 163(01) (2014), 253–85.

[10] A.V. Fursikov and O. Yu. Imanuvilov, On approximate controllability of the Stokes system, *Annales de la Faculté des Sciences de Toulouse*, II(2) (1993), 205–32.

[11] A.V. Fursikov and O. Yu. Imanuvilov, *Controllability of Evolution Equations*, Lecture Notes Series, 34. Seoul National University, Korea (1996), 1–163.

[12] R. Glowinski, *Numerical Methods for Nonlinear Variational Problems*. Springer Series in Computational Physics, 1983.

[13] R. Glowinski, J.L. Lions, and J. He, *Exact and Approximate Controllability for Distributed Parameter Systems: A Numerical Approach*, Encyclopedia of Mathematics and Its Applications, 117. Cambridge University Press, Cambridge, 2008.

[14] O.Yu. Imanuvilov, Remarks on exact controllability for the Navier–Stokes equations, *ESAIM Control Optim. Cal. Var.*, 6 (2001), 39–72.

[15] O.Yu. Imanuvilov, J.-P. Puel, and M. Yamamoto, Carleman estimates for parabolic equations with nonhomogeneous boundary conditions, *Chin. Ann. Math.* 30B(4), 2009, 333–78.

[16] I. Lasiecka and R.Triggiani, *Control Theory for Partial Differential Equations: Continuous and Approximation Theories. I. Abstract Parabolic Systems*. Encyclopedia of Mathematics and Its Applications, 74. Cambridge University Press, Cambridge, 2000.

[17] J.-L. Lions, *Contrôlabilité Exacte, Perturbations et Stabilisation de Systèmes Distribués*, Recherches en Mathématiques Appliquées, Tomes 1 et 2, Masson, Paris, 1988.

Arnaud Münch and Pablo Pedregal

[18] A. Münch, A variational approach to approximate controls for systems with essential spectrum: Application to membranal arch, *Evolut. Eq. Control Theory*, 2(1) (2013), 119–51.

[19] A. Münch, A least-squares formulation for the approximation of controls for the Stokes system. *Math. Controls Signals Syst.*, 27 (2015), 49–75.

[20] A. Münch and E. Zuazua, Numerical approximation of null controls for the heat equation: Ill-posedness and remedies, *Inverse Problems*, 26(8) (2010), 085018, 39 pp.

[21] A. Münch and P. Pedregal, Numerical null controllability of the heat equation through a least squares and variational approach, *Eur. J. Appl. Math.*, 25(03) (2014), 277–306.

[22] A. Münch and P. Pedregal, *A Least-Squares Formulation for the Approximation of Null Controls for the Stokes system*, C.R. Acad. Sci. Série, 1, 351 (2013), 545–50.

[23] A. Münch and P. Pedregal, A least-squares formulation for the approximation of null controls for the Navier–Stokes system. In preparation.

[24] P. Pedregal, A variational perspective on controllability, *Inverse Problems*, 26(1) (2010), 015004, 17 pp.

[25] P. Pedregal, A variational approach for the Navier–Stokes system, *J. Math. Fluid Mech.*, 14(1) (2012), 159–76.

[26] P. Pedregal, On error functionals, *SēMA J.*, 65 (2014), 13–22.

[27] R. Temam, *Navier–Stokes Equations. Theory and Numerical Analysis*. Reprint of the 1984 edition. AMS Chelsea Publishing, Providence, RI, 2001, xiv+408 pp.

[28] D.L. Russell, Controllability and stabilizability theory for linear partial differential equations. Recent progress and open questions, *SIAM Rev.*, 20 (1978), 639–739.

LABORATOIRE DE MATHÉMATIQUES, UNIVERSITÉ BLAISE PASCAL (CLERMONT-FERRAND 2), UMR CNRS 6620, CAMPUS DES CÉZEAUX, 63177 AUBIÈRE, FRANCE
E-mail address: arnaud.munch@math.univ-bpclermont.fr

E.T.S. INGENIEROS INDUSTRIALES. UNIVERSIDAD DE CASTILLA LA MANCHA. CAMPUS DE CIUDAD REAL (SPAIN)
E-mail address: pablo.pedregal@uclm.es

6

A Note on the Asymptotic Stability of Wave-Type Equations with Switching Time-Delay

SERGE NICAISE AND CRISTINA PIGNOTTI

Abstract

We consider second-order evolution equations with intermittently delayed/ not-delayed damping. We give sufficient conditions for asymptotic and exponential stability, completing our previous results from Nicaise and Pignotti [*Adv. Diff. Eq.*, 17: 879–902, 2012; *J. Dyn. Diff. Eq.*, 26: 781–803, 2014]. Moreover, some concrete models are described.

Mathematics Subject Classification 2000. 35L05, 93D15
Key words and phrases. Wave equation, stabilization, switching time delay type feedbacks

Contents

6.1 Introduction

Let H be a real Hilbert space and let $A : \mathcal{D}(A) \to H$ be a positive self-adjoint operator with a compact inverse in H. Denote by $V := \mathcal{D}(A^{\frac{1}{2}})$ the domain of $A^{\frac{1}{2}}$. Let B_1, B_2 be time-dependent linear operators, $B_1 \in \mathcal{L}(H, H)$ and $B_2(t) : U \to H$, where U is a real Hilbert space with norm and inner product denoted respectively by $\| \cdot \|_U$ and $\langle \cdot, \cdot \rangle_U$. We assume that B_1 and B_2 act alternately, that is

$$B_1^*(t) B_2^*(t) = 0, \quad \forall t > 0.$$

Let us consider the problem

$$u_{tt}(t) + Au(t) + B_1(t) B_1^*(t) u_t(t) + B_2(t) B_2^*(t) u_t(t - \tau) = 0 \quad t > 0 \quad (6.1)$$

$$u(0) = u_0 \quad \text{and} \quad u_t(0) = u_1 \tag{6.2}$$

where the constant $\tau > 0$ is the time delay.

Time delay effects appear in many applications and practical problems and it is by now well known that even an arbitrarily small delay in the feedback may destabilize a system which is uniformly exponentially stable in absence of delay. For some examples of this destabilizing effect of time delays we refer to [2, 3, 9, 14].

In [9] we considered the wave equation with both dampings acting simultaneously, that is $B_1(t) = \mu_1$ and $B_2(t) = \mu_2$, with $\mu_1, \mu_2 \in \mathbb{R}^+$, and we proved that if $\mu_1 > \mu_2$ then the system is uniformly exponentially stable. Otherwise, if $\mu_2 \geq \mu_1$, that is the delay term prevails on the not-delayed one, then there are instability phenomena, namely, there are unstable solutions for arbitrarily small (large) delays.

The stabilization problem for second-order evolution equations with switching time delay is, in some sense, related to the one for second-order evolution equations damped by positive/negative feedbacks (see [6]). See also [13] for the relation between wave equation with time delay in the damping and wave equation with indefinite damping, i.e., damping which changes sign in different subsets of the domain.

We firstly studied this subject in [10], in a more general setting with respect to the problem considered here. Assuming that an observability inequality holds for the conservative model associated with (6.1) and (6.2) and, through the definition of a suitable energy, we obtained sufficient conditions ensuring asymptotic stability. Under more restrictive assumptions exponential stability estimates were also obtained.

An analogous problem has been considered in [1] for 1D models for the wave equation but with a different approach. Indeed, in [1] we obtain stability results for particular values of the time delays, related to the length of the domain, by using the D'Alembert formula.

The results of [10] are improved in [11] by removing a quite restrictive assumption on the size of the "bad" terms, i.e., the terms with time delay (cf. assumption (3.3) of [10]). Indeed, expected from [6] and from the relation between delay problems and problems with antidamping, the delay feedback operator B_2 may be also large but then has to act on small time intervals. Moreover, in [11] we consider also the case when B_1 is unbounded. In this short note, we complete the results from [10, 11] by improving the sufficient conditions that guarantee stability results but by staying in the case of bounded operators B_1.

This chapter is organized as follows. In Section 6.2, we recall the well-posedness result proved in [10] and the observability estimate in short-time proved by Haraux, Martinez, and Vancostenoble. In Section 6.3, we obtain asymptotic and exponential stability results for the abstract model under suitable conditions, completing the results of [11]. Finally, in Section 6.4, we give some concrete applications of our results.

6.2 Well-Posedness

In this section we recall a well-posedness result, proved by the authors in [10], for problem (6.1) and (6.2).

We assume that for all $n \in \mathbb{N}$, there exists $t_n > 0$ with $t_n < t_{n+1}$ and such that

$$B_2(t) = 0 \quad \forall t \in I_{2n} = [t_{2n}, t_{2n+1}),$$
$$B_1(t) = 0 \quad \forall t \in I_{2n+1} = [t_{2n+1}, t_{2n+2}),$$

with $B_1 \in C^1([t_{2n}, t_{2n+1}]; \mathcal{L}(H, H))$ and $B_2 \in C([t_{2n+1}, t_{2n+2}]; \mathcal{L}(U, H))$.

Moreover, we assume that $\tau \leq T_{2n}$ for all $n \in \mathbb{N}$, where T_n denotes the length of the interval I_n, i.e.,

$$T_n = t_{n+1} - t_n, \quad n \in \mathbb{N}. \tag{6.3}$$

Under these assumptions, the following result holds (see Theorem 2.1 in [10]).

Theorem 6.1 *Under the above assumptions, for any $u_0 \in V$ and $u_1 \in H$, the system (6.1) and (6.2) has a unique solution $u \in C([0, \infty); V) \cap C^1([0, \infty); H)$.*

To obtain stability results for problem (6.1) and (6.2), we assume now that for all $n \in \mathbb{N}$, there exist three positive constants m_{2n}, M_{2n}, and M_{2n+1} with $m_{2n} \leq M_{2n}$ and such that for all $u \in H$ we have

(i) $m_{2n}\|u\|_H^2 \leq \|B_1^*(t)u\|_H^2 \leq M_{2n}\|u\|_H^2$ for $t \in I_{2n} = [t_{2n}, t_{2n+1})$, $\forall n \in \mathbb{N}$;

(ii) $\|B_2^*(t)u\|_U^2 \leq M_{2n+1}\|u\|_H^2$ for $t \in I_{2n+1} = [t_{2n+1}, t_{2n+2})$, $\forall n \in \mathbb{N}$.

Denote by $E_S(\cdot)$ the standard energy for wave–type equations, i.e.,

$$E_S(t) = E_S(u; t) := \frac{1}{2}\left(\|u(t)\|_V^2 + \|u_t(t)\|_H^2\right).$$

Observe that, on the time intervals I_{2n}, $n \in \mathbb{N}$, only the standard dissipative damping acts. So, the following observability estimate holds (see theorem 3.1 of [6]).

Proposition 6.2 Assume (i). There exists a constant c, independent of the length T_{2n} of the interval I_{2n}, such that for any solution of (6.1) and (6.2) it holds

$$E_S(t_{2n+1}) \leq \frac{1}{1 + c\frac{m_{2n}}{T_{2n}^{-3} + T_{2n}^{-1} + M_{2n}m_{2n}T_{2n}^{-1}}} E_S(t_{2n}), \quad n \in \mathbb{N}. \tag{6.4}$$

6.3 Stability Results

Now, as in [11], we assume

$$\inf_{n \in \mathbb{N}} \frac{m_{2n}}{M_{2n+1}} > 0. \tag{6.5}$$

Note that assumption (3.3) in [10] is instead equivalent to

$$\inf_{n \in \mathbb{N}} \frac{m_{2n}}{M_{2n+1}} > \frac{1}{2}.$$

Let us introduce the energy of the system

$$E(t) = E(u; t) := \frac{1}{2}\left(\|u(t)\|_V^2 + \|u_t(t)\|_H^2 + \frac{\xi}{2}\int_{t-\tau}^{t} \|B_2^*(s+\tau)u_t(s)\|_U^2 \, ds\right), \tag{6.6}$$

where ξ is a positive number satisfying

$$\xi < \inf_{n \in \mathbb{N}} \frac{m_{2n}}{M_{2n+1}}. \tag{6.7}$$

The following estimates are proved as in proposition 3.1 of [11].

Proposition 6.3 Assume (*i*), (*ii*), and (6.5). For any regular solution of problem (6.1) and (6.2) the energy is decreasing on the intervals I_{2n}, $n \in \mathbb{N}$, and

$$E'(t) \leq -\frac{m_{2n}}{2} \|u_t\|_H^2. \tag{6.8}$$

Moreover, on the intervals I_{2n+1}, $n \in \mathbb{N}$,

$$E'(t) \leq \frac{M_{2n+1}}{2} \left(\xi + \frac{1}{\xi} \right) \|u_t\|_H^2. \tag{6.9}$$

The following theorem, due to the more general assumption (6.5) and due to the more general setting, improves and generalizes theorem 3.3 of [10].

Theorem 6.4 *Assume* (*i*), (*ii*), *and* (6.5). *If*

$$\sum_{n=0}^{\infty} M_{2n+1} T_{2n+1} < +\infty, \tag{6.10}$$

and

$$\sum_{n=0}^{\infty} \ln \left(\frac{1}{1 + c \frac{m_{2n}}{T_{2n}^{-3} + T_{2n}^{-1} + M_{2n} m_{2n} T_{2n}^{-1}}} + \tau \xi M_{2n+1} \right) = -\infty, \tag{6.11}$$

then system (6.1) *and* (6.2) *is asymptotically stable, that is any solution u of* (6.1) *and* (6.2) *satisfies* $E_S(u,t) \to 0$ *for* $t \to +\infty$.

Proof Note that (6.9) implies

$$E'(t) \leq M_{2n+1} \left(\xi + \frac{1}{\xi} \right) E(t), \quad t \in I_{2n+1} = [t_{2n+1}, t_{2n+2}), \quad n \in \mathbb{N}.$$

Then we have

$$E(t_{2n+2}) \leq e^{(\xi + \frac{1}{\xi}) M_{2n+1} T_{2n+1}} E(t_{2n+1}), \quad \forall n \in \mathbb{N}. \tag{6.12}$$

Now, note that

$$E(t_{2n+1}) = E_S(t_{2n+1}) + \frac{\xi}{2} \int_{t_{2n+1}-\tau}^{t_{2n+1}} \|B_2^*(s+\tau)u_t(s)\|_U^2 \, ds.$$

Now observe that, since $T_{2n} \geq \tau, n \in \mathbb{N}$, the variable $s+\tau$ in the above integral belongs to $I_{2n+1} \cup I_{2n+2}$. If $s + \tau \in I_{2n+1}$, then $\|B_2^*(s+\tau)u_t(s)\|_U^2 \leq M_{2n+1} \|u_t(s)\|_H^2$. Otherwise, if $s + \tau \in I_{2n+2}$, then $B_2^*(s+\tau)u_t(s) = 0$. Then,

$$E(t_{2n+1}) \leq E_S(t_{2n+1}) + \frac{\xi}{2} M_{2n+1} \int_{t_{2n+1}-\tau}^{t_{2n+1}} \|u_t(s)\|_H^2 \, ds. \tag{6.13}$$

Now, since $(t_{2n+1} - \tau, t_{2n+1}) \subset I_{2n}$ and in I_{2n} the system is dissipative, from (6.13) we deduce

$$E(t_{2n+1}) \leq E_S(t_{2n+1}) + \tau M_{2n+1}\xi E_S(t_{2n+1} - \tau)$$
$$\leq E_S(t_{2n+1}) + \tau M_{2n+1}\xi E_S(t_{2n}). \tag{6.14}$$

Combining Proposition 6.2 and (6.14) we then obtain

$$E(t_{2n+1}) \leq \left(\frac{1}{1 + c\frac{m_{2n}}{T_{2n}^{-3} + T_{2n}^{-1} + M_{2n}m_{2n}T_{2n}^{-1}}} + \tau\xi M_{2n+1} \right) E_S(t_{2n}), \tag{6.15}$$

and therefore

$$E_S(t_{2n+2}) \leq E(t_{2n+2})$$
$$\leq e^{(\xi+\frac{1}{\xi})M_{2n+1}T_{2n+1}} \left(\frac{1}{1 + c\frac{m_{2n}}{T_{2n}^{-3} + T_{2n}^{-1} + M_{2n}m_{2n}T_{2n}^{-1}}} + \tau\xi M_{2n+1} \right) E_S(t_{2n}). \tag{6.16}$$

Since (6.16) holds for any $n \in \mathbb{N}$ we can deduce

$$E_S(t_{2n+2}) \leq \prod_{p=0}^{n} e^{(\xi+\frac{1}{\xi})M_{2p+1}T_{2p+1}} \left(\frac{1}{1 + c\frac{m_{2p}}{T_{2p}^{-3} + T_{2p}^{-1} + M_{2p}m_{2p}T_{2p}^{-1}}} + \tau\xi M_{2p+1} \right) E_S(0). \tag{6.17}$$

Now observe that the standard energy $E_S(\cdot)$ is not decreasing. However, for $t \in [t_{2n}, t_{2n+1})$, only the standard dissipative damping acts and so

$$E_S(t) \leq E_S(t_{2n}), \quad \forall t \in [t_{2n}, t_{2n+1}). \tag{6.18}$$

Moreover, for $t \in [t_{2n+1}, t_{2n+2})$, it results

$$E_S(t) \leq E(t) \leq e^{(\xi+\frac{1}{\xi})M_{2n+1}T_{2n+1}} E(t_{2n+1}), \tag{6.19}$$

where in the second inequality we have used (6.12).

Then, by (6.15) and (6.17)–(6.19), asymptotic stability occurs if

$$\sum_{n=0}^{\infty} \left[\left(\xi + \frac{1}{\xi}\right) M_{2n+1}T_{2n+1} + \ln \tilde{c}_n \right] = -\infty, \tag{6.20}$$

where

$$\tilde{c}_n = \frac{1}{1 + c\frac{m_{2n}}{T_{2n}^{-3} + T_{2n}^{-1} + M_{2n}m_{2n}T_{2n}^{-1}}} + \tau\xi M_{2n+1}. \tag{6.21}$$

Thus, if

$$\sum_{n=0}^{\infty} M_{2n+1} T_{2n+1} < +\infty, \qquad \sum_{n=0}^{\infty} \ln \tilde{c}_n = -\infty,$$

system (6.1) and (6.2) is asymptotically stable.   □

We now show that under additional assumptions on the coefficients T_n, m_n, M_n an exponential stability result holds.

Theorem 6.5 *Assume* (i), (ii), *and* (6.5). *Assume also that*

$$T_{2n} = T^* \quad \forall n \in \mathbb{N}, \tag{6.22}$$

with $T^* \geq \tau$, *and*

$$T_{2n+1} = \tilde{T} \quad \forall n \in \mathbb{N}. \tag{6.23}$$

Moreover, assume that

$$\sup_{n \in \mathbb{N}} e^{(\xi + \frac{1}{\xi}) M_{2n+1} \tilde{T}} \tilde{c}_n = d < 1, \tag{6.24}$$

where $\tilde{c}_n$ *is as in* (6.21). *Then, there exist two positive constants* γ, μ *such that*

$$E_S(t) \leq \gamma e^{-\mu t} E_S(0), \quad t > 0, \tag{6.25}$$

for every solution of problem (6.1) − (6.2).

Proof From (6.16) and (6.24) we obtain

$$E_S(T^* + \tilde{T}) \leq d E_S(0),$$

and also

$$E_S(n(T^* + \tilde{T})) \leq d^n E_S(0), \quad \forall n \in \mathbb{N}.$$

Then, the standard energy $E_S(\cdot)$ satisfies an exponential estimate like (6.25) (see lemma 1 of [5]).   □

Remark 6.6 In the assumptions of Theorem 6.5, from (6.17) we can see that exponential stability also holds if instead of (6.24) we assume

$$\exists n \in \mathbb{N} \quad \text{such that} \quad \prod_{p=k(n+1)}^{k(n+1)+n} e^{(\xi + \frac{1}{\xi}) M_{2p+1} \tilde{T}} c_p \leq d < 1, \quad \forall k = 0, 1, 2, \ldots$$

6.4 Examples

Here, we illustrate some concrete examples falling in our previous abstract setting, namely the wave equation, the elasticity system, the Midlin–Timoshenko model, the Petrowsky system.

6.4.1 The Wave Equation

As a concrete application let us consider the wave equation with internal damping. More precisely, let $\Omega \subset \mathbb{R}^d$ (with d a positive natural number) be an open bounded domain with a boundary $\partial\Omega$ of class C^2. We denote by ω a subset of Ω.

Let us consider the initial boundary value problem

$$u_{tt}(x,t) - \Delta u(x,t) + b_1(t)u_t(x,t) + b_2(t)\chi_\omega u_t(x,t-\tau) = 0$$
$$\text{in } \Omega \times (0,+\infty) \qquad (6.26)$$

$$u(x,t) = 0 \quad \text{on } \partial\Omega \times (0,+\infty) \qquad (6.27)$$

$$u(x,0) = u_0(x) \quad \text{and} \quad u_t(x,0) = u_1(x) \quad \text{in } \Omega \qquad (6.28)$$

with initial data $(u_0, u_1) \in H_0^1(\Omega) \times L^2(\Omega)$ and b_1, b_2 in $L^\infty(0,+\infty)$ such that

$$b_1(t)b_2(t) = 0, \quad \forall t > 0.$$

Moreover, we assume

(i_w) $0 < m_{2n} \leq b_1(t) \leq M_{2n}$, $b_2(t) = 0$, for all $t \in I_{2n} = [t_{2n}, t_{2n+1})$, and $b_1 \in C^1(\bar{I}_{2n})$, for all $n \in \mathbb{N}$;

(ii_w) $|b_2(t)| \leq M_{2n+1}$, $b_1(t) = 0$, for all $t \in I_{2n+1} = [t_{2n+1}, t_{2n+2})$, and $b_2 \in C(\bar{I}_{2n+1})$, for all $n \in \mathbb{N}$.

This problem enters into our previous framework, if we take $H = L^2(\Omega)$ and the operator A defined by

$$A : \mathcal{D}(A) \to H : u \to -\Delta u,$$

where $\mathcal{D}(A) = H_0^1(\Omega) \cap H^2(\Omega)$.

The operator A is a self-adjoint and positive operator with a compact inverse in H and is such that $V = \mathcal{D}(A^{1/2}) = H_0^1(\Omega)$. We then define the operator B_1 as

$$B_1 : H \to H : v \to \sqrt{b_1}v, \qquad (6.29)$$

and, denote $U := L^2(\Omega)$, the operator B_2 as

$$B_2 : U \to H : v \to \sqrt{b_2}\tilde{v}, \qquad (6.30)$$

where $\tilde{v} \in L^2(\Omega)$ is the extension of v by zero outside ω.

It is easy to verify that $B_1 B_1^*(\varphi) = b_1\varphi$ and $B_2 B_2^*(\varphi) = b_2\varphi\chi_\omega$, for $\varphi \in H$.

This shows that problem (6.26) and (6.28) enters in the abstract framework (6.1) and (6.2). Moreover, (i_w) and (ii_w) easily imply (i) and (ii) of Section 6.3. Therefore we can restate Proposition 6.3. Now, the energy functional is

$$E(t) = \frac{1}{2} \int_\Omega \{u_t^2(x, t) + |\nabla u(x, t)|^2\}dx + \frac{\xi}{2} \int_{t-\tau}^t |b_2(s + \tau)| \int_\omega u_t^2(x, s)dx\,ds.$$
$$(6.31)$$

Proposition 6.7 Assume (i_w), (ii_w), and (6.5). Then, for every regular solution of problem (6.26)–(6.28) the energy is decreasing on the intervals I_{2n}, $n \in \mathbb{IN}$, and

$$E'(t) \leq -\frac{m_{2n}}{2} \int_\Omega u_t^2(x, t)dx. \qquad (6.32)$$

Moreover, on the intervals I_{2n+1}, $n \in \mathbb{IN}$,

$$E'(t) \leq \frac{M_{2n+1}}{2} \left(\xi + \frac{1}{\xi}\right) \int_\Omega u_t^2(x, t)dx. \qquad (6.33)$$

Thus, the stability results of Theorems 6.4 and 6.5 apply to this model.

6.4.2 The Elasticity System

In the same setting than in the previous section, we consider the following elastodynamic system

$$u_{tt}(x, t) - \mu\Delta u(x, t) - (\lambda + \mu)\nabla \operatorname{div} u(x, t) + b_1(t)u_t(x, t)$$
$$+ b_2(t)\chi_\omega u_t(x, t - \tau) = 0 \quad \text{in } \Omega \times (0, +\infty) \qquad (6.34)$$
$$u(x, t) = 0 \quad \text{on } \partial\Omega \times (0, +\infty) \qquad (6.35)$$
$$u(x, 0) = u_0(x) \quad \text{and} \quad u_t(x, 0) = u_1(x) \quad \text{in } \Omega \qquad (6.36)$$

with initial data $(u_0, u_1) \in H_0^1(\Omega)^d \times L^2(\Omega)^d$ and b_1, b_2 satisfying the same assumptions as in Section 6.4.1. Here the state variable u is vector-valued and λ, μ are the Lamé coefficients that are positive real numbers. Finally for a (smooth enough) vector-valued function $v : \Omega \to \mathbb{R}^d$, div v is its standard divergence, namely

$$\operatorname{div} v = \sum_{j=1}^d \partial_j v_j.$$

As before this problem enters into our abstract setting, once we take $H = L^2(\Omega)^d$, and A defined by

$$A : \mathcal{D}(A) \to H : u \to -\mu \Delta u(x, t) - (\lambda + \mu)\nabla \operatorname{div} u,$$

where $\mathcal{D}(A) = H_0^1(\Omega)^d \cap H^2(\Omega)^d$.

The operator A is a self-adjoint and positive operator with a compact inverse in H and is such that $V = \mathcal{D}(A^{1/2}) = H_0^1(\Omega)^d$ equipped with the inner product

$$(u, v)_V = \int_\Omega \left(\mu \sum_{i,j=1}^d \partial_i u_j \partial_i v_j + (\lambda + \mu) \operatorname{div} u \operatorname{div} v \right) dx, \quad \forall u, v \in H_0^1(\Omega)^d.$$

We further define $U = L^2(\omega)^d$ and the operators B_i, $i = 1, 2$, as follows:

$$B_1 : U \to H : v \to \sqrt{b_1} v,$$
$$B_2 : U \to H : v \to \sqrt{b_2} \tilde{v},$$

$\tilde{v}$ being again the extension of v by zero outside ω. As before $B_1 B_1^*(\varphi) = b_1 \varphi$, and $B_2 B_2^*(\varphi) = b_2 \varphi \chi_\omega$, for any $\varphi \in H$. So, problem (6.34)–(6.36) enters in the abstract framework (6.1) and (6.2). Moreover, (i_w) and (ii_w) easily imply (i) and (ii) of Section 6.2. Therefore, the results of Section 6.3 apply also to the system (6.34)–(6.36) with the energy defined by

$$E(t) = \frac{1}{2} \int_\Omega \left\{ |u_t|^2(x, t) + \mu \sum_{i,j=1}^d (\partial_i u_j(x, t))^2 + (\lambda + \mu)(\operatorname{div} u(x, t))^2 \right\} dx$$

$$+ \frac{\xi}{2} \int_{t-\tau}^t |b_2(s + \tau)| \int_\omega |u_t(x, s)|^2 dx\, ds.$$

6.4.3 The Mindlin–Timoshenko Model

In the same setting as in Section 6.4.1, we consider the internal stabilization of the following Mindlin–Timoshenko (beam/plate) model (for similar models, see chapter 5 of [8], chapters 2 and 4 of [7], [4, 12]).

$$w_{tt}(x, t) = \operatorname{div}(K(\nabla w + u))(x, t) - a_1(t)w_t(x, t)$$
$$- a_2(t)\chi_\omega w_t(x, t - \tau) \quad \text{in } \Omega \times (0, +\infty) \tag{6.37}$$

$$u_{tt}(x, t) = \operatorname{div} C\epsilon(u)(x, t) - K(\nabla w + u)(x, t) - b_1(t)u_t(x, t)$$
$$- b_2(t)\chi_\omega u_t(x, t - \tau) \quad \text{in } \Omega \times (0, +\infty) \tag{6.38}$$

$$u(x,t) = 0, \quad w(x,t) = 0 \quad \text{on } \partial\Omega \times (0, +\infty) \tag{6.39}$$

$$u(x,0) = u^0(x), \, u_t(x,0) = u^1(x), \, w(x,0) = w^0(x), \, w_t(x,0) = w^1(x) \text{ in } \Omega. \tag{6.40}$$

If $d=1$ (resp. $d=2$) the scalar variable w represents the displacement of the beam (resp. plate) in the vertical direction, while the vectorial variable $u = (u_i)_{i=1}^d$ is the angles of rotation of a filament of the beam (resp. plate).

Here K belongs to $L^\infty(\Omega)^{d \times d}$, is symmetric and positive definite, i.e., there exists a positive constant k_0 such that

$$X^\top K(x) X \geq k_0, \quad \forall X \in \mathbb{R}^d, \text{ for a.e. } x \in \Omega,$$

while $C = (c_{ijk\ell})$ is a tensor such that

$$c_{ijk\ell} = c_{ji\ell k} = c_{k\ell ij} \in L^\infty(\Omega), \tag{6.41}$$

all indices running over the integers $1, \ldots, d$. These quantities are related to the constitutive materials of the beam/plate.

As usual for $u = (u_i)_{i=1}^d$, $\epsilon(u)$ is the linear strain tensor defined by

$$\epsilon(u) = (\epsilon_{ij}(u))_{i,j=1}^d \quad \text{with } \epsilon_{ij}(u) = \frac{1}{2}(\partial_i u_j + \partial_j u_i).$$

For a $d \times d$ matrix $\epsilon = (\epsilon_{ij})_{i,j=1}^d$ the product $C\epsilon = ((C\epsilon)_{ij})_{i,j=1}^d$ is the $d \times d$ matrix given by

$$(C\epsilon)_{ij} = \sum_{k,\ell=1}^d c_{ijk\ell}\epsilon_{k\ell}.$$

As usual we assume that C is positive definite in the sense that there exists $\mu_0 > 0$ such that

$$C(x)\epsilon : \epsilon \geq \mu_0 |\epsilon|^2, \quad \forall \epsilon \in \mathbb{R}^{d \times d}, \text{ for a.e. } x \in \Omega. \tag{6.42}$$

We further recall that for a (smooth enough) matrix-valued function $w = (w_{ij}) : \Omega \to \mathbb{R}^{d \times d}$, div w is its divergence line by line, i.e.,

$$\text{div } w = \left(\sum_{j=1}^d \partial_j w_{ij} \right)_{i=1}^d.$$

Finally we require that the functions b_1, b_2 satisfy the assumptions (i_w) and (ii_w) from Section 6.4.1, and similarly for a_1 and a_2, we suppose that

(i_{mt}) $0 < m_{2n} \leq a_1(t) \leq M_{2n}$, $a_2(t) = 0$, for all $t \in I_{2n} = [t_{2n}, t_{2n+1})$, and $a_1 \in C^1(\bar{I}_{2n})$, for all $n \in \mathbb{N}$;

(ii_{mt}) $|a_2(t)| \le M_{2n+1}$, $a_1(t) = 0$, for all $t \in I_{2n+1} = [t_{2n+1}, t_{2n+2})$, and $a_2 \in C(\bar{I}_{2n+1})$, for all $n \in \mathbb{N}$.

System (6.37)–(6.39) can be viewed as a coupling between the wave equation in w with the dynamical elastic system in u with internal feedbacks with delays.

Again this problem enters into our abstract setting by using Friedrichs extension theorem. Namely we take $H = L^2(\Omega)^d \times L^2(\Omega)$ with its natural inner product

$$((u, w), (u^*, w^*))_H = \int_\Omega \left(u \cdot \bar{u}^* + w\bar{w}^* \right) dx, \quad \forall (u, w), (u^*, w^*) \in H,$$

$V = H_0^1(\Omega)^d \times H_0^1(\Omega)$ that is clearly compactly embedded into H and the sesquilinear and symmetric form

$$a(U, U^*) = \int_\Omega \left(C\varepsilon(u) : \varepsilon(\bar{u}^*) + K(\nabla w + u) \cdot (\nabla \bar{w}^* + \bar{u}^*) \right) dx$$

with $U = (u, w)$, $U^* = (u^*, w^*) \in V$. Indeed using Korn's and Poincaré's inequalities, it is not difficult to check that this sesquilinear form is coercive in V, namely there exists $\alpha > 0$ such that

$$a(U, U) \ge \alpha \|U\|_{H^1(\Omega)^d \times H^1(\Omega)}^2, \quad \forall U \in V.$$

Hence it is well known that the operator A associated with the triple (a, V, H) is a self-adjoint and positive operator with a compact inverse in H with $\mathcal{D}(A^{1/2}) = V$. This operator is defined by

$$\mathcal{D}(A) = \{U \in V : \exists F_U \in H \text{ such that } a(U, U^*) = (F_H, U^*)_H, \forall U^* \in V\},$$

and

$$AU = F_U, \quad \forall U \in \mathcal{D}(A).$$

By the definition of a, it is easy to see that

$$\mathcal{D}(A) = \{(u, w) \in V : \text{div } C\varepsilon(u) \in L^2(\Omega)^d \text{ and div}(K(\nabla w + u)) \in L^2(\Omega)\},$$

and then

$$A(u, w) = -(\text{div } C\varepsilon(u), \text{div}(K(\nabla w + u))), \quad \forall (u, w) \in \mathcal{D}(A).$$

We further define $U = H$ and the operators B_i, $i = 1, 2$, as follows:

$$B_1 : (u, w) \to H : v \to (\sqrt{b_1}v, \sqrt{a_1}w)$$
$$B_2 : (u, w) \to H : v \to (\sqrt{b_2}\tilde{v}, \sqrt{a_2}\tilde{w}),$$

$\tilde{v}$ being again the extension of v by zero outside ω. Clearly one has $B_1 B_1^*(u, w) = (b_1 u, a_1 w)$ and $B_2 B_2^*(u, w) = (b_2 u \chi_\omega, a_2 w \chi_\omega)$, for any $(u, w) \in H$. So, problem (6.37)–(6.39) enters in the abstract framework (6.1) and (6.2). Moreover, (i_w), (ii_w), (i_{mt}), and (ii_{mt}) imply (i) and (ii) from Section 6.2, and consequently the results of Section 6.3 apply also to this system with the energy defined by

$$E(t) = \frac{1}{2} \int_\Omega \left\{ (C\varepsilon(u)(x, t)) : \varepsilon(\bar{u}(x, t))) + K(\nabla w + u)(x, t)) \cdot (\nabla \bar{w} + \bar{u})(x, t)) \right\} dx$$
$$+ \frac{\xi}{2} \int_{t-\tau}^{t} \int_\omega (|b_2(s + \tau)| |u_t(x, s)|^2 + |a_2(s + \tau)| |w_t(x, s)|^2) dx \, ds.$$

6.4.4 The Petrovsky System

Let $\Omega \subset \mathbb{R}^d$ be an open bounded set with a boundary $\partial \Omega$ of class C^4 and let ω be any fixed subset of Ω.

Let us consider the initial boundary value problem

$$u_{tt}(x, t) + \Delta^2 u(x, t) + b_1(t) u_t(x, t) + b_2(t) \chi_\omega u_t(x, t - \tau) = 0$$
$$\text{in } \Omega \times (0, +\infty) \qquad (6.43)$$

$$u(x, t) = \Delta u(x, t) = 0 \quad \text{on } \partial \Omega \times (0, +\infty) \qquad (6.44)$$

$$u(x, 0) = u_0(x) \quad \text{and} \quad u_t(x, 0) = u_1(x) \quad \text{in } \Omega \qquad (6.45)$$

with initial data $(u_0, u_1) \in H^2(\Omega) \cap H_0^1(\Omega) \times L^2(\Omega)$ and b_1, b_2 satisfying the same assumptions as in Section 6.4.1.

Now, we take $H = L^2(\Omega)$ and let A be the operator

$$A : \mathcal{D}(A) \to H : u \to \Delta^2 u, \qquad (6.46)$$

where

$$\mathcal{D}(A) = \{v \in H_0^1(\Omega) \cap H^4(\Omega) : \Delta u = 0 \text{ on } \partial \Omega\}.$$

The operator A is self-adjoint and positive, has a compact inverse in H and satisfies $\mathcal{D}(A^{1/2}) = H^2(\Omega) \cap H_0^1(\Omega)$. We then define $U = L^2(\omega)$ and the operators B_i, $i = 1, 2$, by (6.29) and (6.30). So, problem (6.43)–(6.45) enters in the abstract framework (6.1) and (6.2). Moreover, (i_w) and (ii_w) easily imply (i) and (ii) of Section 6.2.

Therefore, the results of Section 6.3 apply also to the plate model.

References

[1] K. Ammari, S. Nicaise, and C. Pignotti. Stabilization by switching time-delay. *Asymptot. Anal.*, 83:263–83, 2013.

[2] R. Datko. Not all feedback stabilized hyperbolic systems are robust with respect to small time delays in their feedbacks. *SIAM J. Control Optim.*, 26:697–713, 1988.

[3] R. Datko, J. Lagnese, and M. P. Polis. An example on the effect of time delays in boundary feedback stabilization of wave equations. *SIAM J. Control Optim.*, 24:152–6, 1986.

[4] H. D. Fernández Sare, On the stability of Mindlin–Timoshenko plates. *Quart. Appl. Math.*, 67:249–63, 2009.

[5] M. Gugat. Boundary feedback stabilization by time delay for one-dimensional wave equations. *IMA J. Math. Control Inform.*, 27:189–203, 2010.

[6] A. Haraux, P. Martinez, and J. Vancostenoble. Asymptotic stability for intermittently controlled second-order evolution equations. *SIAM J. Control Optim.*, 43:2089-108, 2005.

[7] J. Lagnese and J.-L. Lions, Modelling analysis and control of thin plates, *Recherches en Mathématiques Appliquées [Research in Applied Mathematics]*, 6, Masson, Paris, 1988.

[8] J. E. Lagnese, Boundary stabilization of thin plates, *SIAM Studies in Applied Mathematics*, 10, Society for Industrial and Applied Mathematics (SIAM), Philadelphia, PA, 1989.

[9] S. Nicaise and C. Pignotti. Stability and instability results of the wave equation with a delay term in the boundary or internal feedbacks. *SIAM J. Control Optim.*, 45:1561–85, 2006.

[10] S. Nicaise and C. Pignotti. Asymptotic stability of second-order evolution equations with intermittent delay. *Adv. Diff. Eq.*, 17: 879–902, 2012.

[11] S. Nicaise and C. Pignotti. Stability results for second-order evolution equations with switching time-delay. *J. Dyn. Diff. Eq.*, 26:781–803, 2014.

[12] S. Nicaise, Internal stabilization of a Mindlin–Timoshenko model by interior feedbacks. *Math. Control Relat. Fields*, 1:331–52, 2011.

[13] C. Pignotti. A note on stabilization of locally damped wave equations with time delay. *Systems Control Lett.*, 61:92–7, 2012.

[14] G. Q. Xu, S. P. Yung, and L. K. Li. Stabilization of wave systems with input delay in the boundary control. *ESAIM: Control Optim. Calc. Var.*, 12:770–85, 2006.

Université de Valenciennes et du Hainaut Cambrésis, Laboratoire de Mathématiques et leurs Applications, Institut des Sciences et Techniques de Valenciennes, 59313 Valenciennes Cedex 9, France

E-mail address: snicaise@univ-valenciennes.fr

Università di L'Aquila, Dipartimento di Ingegneria e Scienze dell'Informazione e Matematica, Via Vetoio, Loc. Coppito, 67010 L'Aquila Italy

E-mail address: pignotti@univaq.it

7

Ill-Posedness of Coupled Systems with Delay

LISA FISCHER AND REINHARD RACKE

Abstract

We consider linear initial-boundary value problems that are a coupling like second-order thermoelasticity, or the thermoelastic plate equation or its generalization (the α–β-system). Now, there is a delay term given in part of the coupled system, and we demonstrate that the expected inherent damping will not prevent the system from not being stable; indeed, the systems will be shown to be ill-posed: a sequence of bounded initial data may lead to exploding solutions (at any fixed time). This is extended to systems with Kelvin–Voigt-like damping terms.

1991 Mathematics Subject Classification. 35B, 35K20, 35K55, 35L20, 35Q

Key words and phrases. Ill-posedness, coupled systems with delay

Contents

7.1 Introduction

Simplest delay equations of parabolic type like

$$\theta_t(t, x) = \Delta\theta(t - \tau), \tag{7.1}$$

or of hyperbolic type like

$$u_{tt}(t, x) = \Delta u(t - \tau),\qquad(7.2)$$

with $\tau > 0$ being a delay parameter, are ill-posed: There is a sequence of initial data remaining bounded, while the corresponding solutions, at a fixed time, go to infinity in an exponential manner.

The same holds for more general, abstract equations like

$$\frac{d^n}{dt^n}u(t) = Au(t - \tau),\qquad(7.3)$$

with $n \in \mathbb{N}$ fixed, $(-A)$ being a linear operator in a Banach space having a sequence of real eigenvalues $(\lambda_k)_k$ such that $0 < \lambda_k \to \infty$ as $k \to \infty$ (see [6, 9]).

Adding certain nondelay terms, e.g., $\Delta\theta(t, x)$ on the right-hand side of (7.1), is – for example – sufficient to obtain a well-posed problem (cp. [10, 25]).

We consider coupled systems of different types:
Hyperbolic–parabolic system,

$$u_{tt}(t, x) - au_{xx}(t - \tau, x) + b\theta_x(t, x) = 0,\qquad(7.4)$$
$$\theta_t(t, x) - d\theta_{xx}(t, x) + bu_{tx}(t, x) = 0,\qquad(7.5)$$

or Schrödinger type coupled to parabolic equation,

$$u_{tt}(t, x) + a\Delta^2 u(t - \tau, x) + b\Delta\theta(t, x) = 0,\qquad(7.6)$$
$$\theta_t(t, x) - d\Delta\theta(t, x) - b\Delta u_t(t, x) = 0,\qquad(7.7)$$

or, more general α–β-systems with delay:

$$u_{tt}(t) + aAu(t - \tau) - bA^\beta\theta(t) = 0,\qquad(7.8)$$
$$\theta_t(t) + dA^\alpha\theta(t) + bA^\beta u_t(t) = 0.\qquad(7.9)$$

$u, \theta : [0, \infty) \to \mathcal{H}$, where $\mathcal{H}$ is a Hilbert space, and A is self-adjoint having a countable complete orthonormal system of eigenfunctions $(\phi_j)_j$ with corresponding eigenvalues $0 < \lambda_j \to \infty$ as $j \to \infty$.

For *well-posedness results for wave equations* with delay terms (in the interior) (see [2, 8, 23, 24], in [23] also with instability results). For the system of *coupled wave equations* of Timoshenko type with delay terms see [12, 28].

The thermoelastic plate system (7.6) and (7.7) has been widely discussed for $\tau = 0$ (see [3, 4, 5, 11, 13, 14, 15, 16, 17, 18, 19, 20, 21, 22, 27]). The more general α–β-system without delay ($\tau = 0$) was introduced in [1, 22].

For the case $\tau = 0$, i.e., without delay, we know the there is a smoothing effect in a certain region (see Figure 7.1).

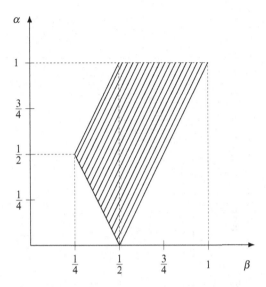

Fig. 7.1.   Area of smoothing $\mathcal{A}_{sm}$ (without delay)

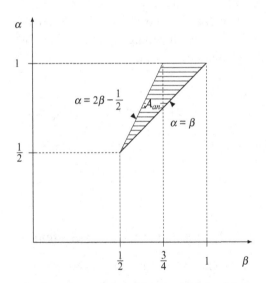

Fig. 7.2.   Area of analyticity $\mathcal{A}_{an}$ (without delay)

Also the region of analyticity is known (see Figure 7.2).

For $\tau > 0$ we can describe now the area of ill-posedness for (7.8) and (7.9) (see Figure 7.3):

$$\mathcal{A}_{ill} := \{(\beta,\alpha) \mid 0 < \beta \leq \alpha \leq 1, \ (\beta,\alpha) \neq (1,1)\}. \tag{7.10}$$

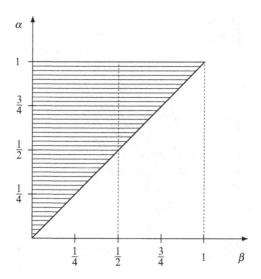

Fig. 7.3.   Area of ill-posedness $\mathcal{A}_{ill}$ (with delay)

Moreover, we shall discuss the α–β–γ-system, where the α–β-system is augmented by an additional damping in the equation for u, representing for $\gamma = 1$ a Kelvin–Voigt type damping:

$$u_{tt}(t) + aAu(t - \tau) - bA^\beta \theta(t) + cA^\gamma u_t(t) = 0, \qquad (7.11)$$
$$\theta_t(t) + dA^\alpha \theta(t) + bA^\beta u_t(t) = 0. \qquad (7.12)$$

The results in Section 7.2 are given in [26], those in Section 7.4 are based on [7].

7.2 Results for the α–β-System

To complete the problem, we have given initial data:

$$u(s) = u^0(s), \quad (-\tau \leq s \leq 0), \quad u_t(0) = u^1, \quad \theta(0) = 0^0. \qquad (7.13)$$

Theorem 7.1 *Let* $(\beta, \alpha) \in \mathcal{A}_{ill}$. *Then the delay problem (7.8), (7.9), and (7.13) is ill-posed. There exists a sequence* $(u_j, \theta_j)_j$ *of solutions with norm* $\|u_j(t)\|_{\mathcal{H}}$ *tending to infinity (as $j \to \infty$) for any fixed t, while for the initial data the norms* $\sup_{-\tau \leq s \leq 0} \|(u_j^0(s), u_j^1, \theta^0)\|_{\mathcal{H}^3}$ *remain bounded.*

7.3 Proof of Theorem 7.1

We make the ansatz

$$u = u_j(t) = h_j(t)\phi_j, \tag{7.14}$$

$$\theta = \theta_j(t) = g_j(t)\phi_j. \tag{7.15}$$

Plugging this into the differential equations, we obtain

$$h_j'''(t) + d\lambda_j^\alpha h_j''(t) + b^2\lambda_j^{2\beta} h_j'(t) + a\lambda_j h_j'(t-\tau) + ad\lambda_j^{1+\alpha} h_j(t-\tau) = 0, \tag{7.16}$$

$$g_j'(t) + d\lambda_j^\alpha g_j(t) = -b\lambda_j^\beta h_j'(t), \tag{7.17}$$

with

$$g_j(0) := \frac{1}{b\lambda_j^\beta}(h_j''(0) + a\lambda_j h_j(-\tau)). \tag{7.18}$$

With the ansatz

$$h_j(t) = \frac{1}{\omega_j^2} e^{\omega_j t}, \tag{7.19}$$

it is then sufficient and necessary that

$$\omega_j^3 + d\lambda_j^\alpha \omega_j^2 + (b^2\lambda_j^{2\beta} + a\lambda_j e^{-\tau\omega_j})\omega_j = -ad\lambda_j^{1+\alpha} e^{-\tau\omega_j} \tag{7.20}$$

holds. We try to find subsequence $(\omega_{j_k})_k$ with

$$\Re\omega_{j_k} \to \infty \quad \text{as } k \to \infty, \tag{7.21}$$

$$\sup_k \left| \frac{\lambda_{j_k}^{1-\beta} e^{-\tau\omega_{j_k}}}{\omega_{j_k}^2} \right| < \infty, \tag{7.22}$$

to assure the boundedness of the data, and

$$\text{for } t > 0: \left| \frac{e^{\omega_{j_k} t}}{\omega_{j_k}^2} \right| \to \infty \quad \text{as } k \to \infty. \tag{7.23}$$

Case 1: $\alpha < 2\beta$.

(7.20) is equivalent to

$$\omega_j\left(1 + \frac{\omega_j^2}{b^2\lambda_j^{2\beta}} + \frac{d\omega_j}{b^2\lambda_j^{2\beta-\alpha}} + \frac{a}{b^2}\lambda_j^{1-2\beta} e^{-\tau\omega_j}\right) = \frac{-ad}{b^2}\lambda_j^{1+\alpha-2\beta} e^{-\tau\omega_j}. \tag{7.24}$$

Let μ_j, ζ_j be defined by

$$\omega_j = \mu_j(1 + \zeta_j), \tag{7.25}$$

where $|\zeta_j| < \frac{1}{2}$ and

$$\mu_j = \frac{-ad}{b^2}\lambda_j^{1+\alpha-2\beta}e^{-\tau\mu_j}. \tag{7.26}$$

(7.26) has a subsequence $(\mu_{j_k})_k$ of solutions with $\Re\mu_{j_k} \to \infty$ (see [6]). (7.24) is equivalent to (omitting indices j or j_k)

$$\underbrace{\left(1 - e^{-\tau\mu\zeta}\right)}_{=:f(\zeta)} + \underbrace{\left(q(\zeta) + \zeta + \zeta q(\zeta)\right)}_{=:g(\zeta)} = 0, \tag{7.27}$$

where

$$q(\zeta) := \frac{\mu^2(1+\zeta)^2}{b^2\lambda^{2\beta}} + \frac{d\mu(1+\zeta)}{b^2\lambda^{2\beta-\alpha}} + \frac{a}{b^2}\lambda^{1-2\beta}e^{-\tau\mu(1+\zeta)}. \tag{7.28}$$

f, g are holomorphic in $B\left(0, \frac{1}{10\tau|\mu|}\right)$ with a single zero of f there.

$$|g(\zeta)| \le \frac{C}{|\mu|}, \tag{7.29}$$

using the information on α, β. Rouché's theorem gives the desired zero ζ.

Case 2: $\alpha \ge 2\beta$.

(7.20) is now equivalent to

$$\omega^2\left(1 + \frac{\omega}{d\lambda^\alpha} + \frac{b^2}{\lambda^{\alpha-2\beta}\omega} + \frac{a\lambda^{1-\alpha}}{d\omega}e^{-\tau\omega}\right) = -a\lambda e^{-\tau\omega}. \tag{7.30}$$

With the ansatz (7.25) again, now with μ solving

$$\mu^2 = -a\lambda e^{-\tau\mu} \tag{7.31}$$

we can proceed as in Case 1.

7.4 Results for the α–β–γ-System

We remark that an additional damping u_t in (7.8) does not lead to well-posedness. An additional damping $A^\gamma u_t$ in (7.8) leads to ill-posedness in regions depending on α, β, γ. Let

$$\mathcal{A}_{ill}(\gamma) := \{(\beta, \alpha, \gamma) \mid 0 < \gamma \le \beta \le \alpha \le 1, (\beta, \alpha) \ne (1, 1), (\alpha \ge 2\beta \vee 2\beta > \alpha + \gamma)\}. \tag{7.32}$$

Theorem 7.2 *Let* $(\beta, \alpha, \gamma) \in \mathcal{A}_{ill}(\gamma)$. *Then the delay problem (7.11) and (7.12) with initial data (7.13) is ill-posed. There exists a sequence* $(u_j, \theta_j)_j$ *of solutions with norm* $\|u_j(t)\|_{\mathcal{H}}$ *tending to infinity (as* $j \to \infty$*) for any fixed* t, *while for the initial data the norms* $\sup_{-\tau \le s \le 0} \|(u_j^0(s), u_j^1, \theta^0)\|_{\mathcal{H}^3}$ *remain bounded.*

7.5 Proof of Theorem 7.2

We make the ansatz (7.14) for $(u_j, \theta_j)_j$ again and look for solutions of the ordinary differential equation

$$h_j'''(t) + d\lambda_j^\alpha h_j''(t) + \left(b^2\lambda_j^{2\beta} + c\lambda_j^\gamma + dc\lambda_j^{\alpha+\gamma}\right)h_j'(t)$$
$$+ a\lambda_j h_j'(t-\tau) + ad\lambda_j^{\alpha+1} h_j(t-\tau) = 0, \tag{7.33}$$

where g_j is determined by

$$g_j'(t) + d\lambda_j^\alpha g_j(t) + b\lambda_j^\beta h_j'(t) = 0 \tag{7.34}$$

with

$$g_j(0) = \frac{1}{b\lambda_j^\beta}\left(h_j''(0) + a\lambda_j h_j(-\tau) + c\lambda_j^\gamma h_j'(0)\right). \tag{7.35}$$

Using the ansatz (7.19) once more it is necessary and sufficient that $(\omega_j)_j$ satisfies

$$\omega_j^3 + d\lambda_j^\alpha \omega_j^2 + \left(b^2\lambda_j^{2\beta} + c\lambda_j^\gamma + dc\lambda_j^{\alpha+\gamma}\right)\omega_j + a\lambda_j\omega_j e^{-\tau\omega_j} + ad\lambda_j^{\alpha+1}e^{-\tau\omega_j} = 0. \tag{7.36}$$

The aim is to find a subsequence $(\omega_{j_k})_k$, which fulfills (7.21)–(7.23).

Case 1: $\alpha \ge 2\beta$.

(7.36) is equivalent to

$$\omega_j\left(1 + \frac{\omega_j^2}{dc\lambda_j^{\alpha+\gamma}} + \frac{\omega_j}{c\lambda_j^\gamma} + \frac{b^2}{dc\lambda_j^{\alpha+\gamma-2\beta}} + \frac{1}{d\lambda_j^\alpha} + \frac{ae^{-\tau\omega_j}}{dc\lambda_j^{\alpha+\gamma-1}}\right) = -\frac{a}{c}\lambda_j^{1-\gamma}e^{-\tau\omega_j}. \tag{7.37}$$

Let μ_j, ζ_j be defined by

$$\omega_j = \mu_j(1 + \zeta_j) \tag{7.38}$$

where $\zeta_j \in \mathbb{C}$, $|\zeta_j| < \frac{1}{2}$ and

$$\mu_j = -\frac{a}{c}\lambda_j^{1-\gamma}e^{-\tau\mu_j}. \tag{7.39}$$

As in the proof of Theorem 7.1 there is a subsequence of solutions to (7.39) $(\mu_{j_k})_k$ with $\Re\mu_{j_k} \longrightarrow \infty$.

Moreover, (7.37) is equivalent to (omitting indices j and j_k)

$$\underbrace{(1 - e^{-\tau\mu\zeta})}_{=:f(\zeta)} + \underbrace{(q(\zeta) + \zeta q(\zeta) + \zeta)}_{=:g(\zeta)} = 0, \qquad (7.40)$$

where

$$q(\zeta) := \frac{\mu^2(1+\zeta)^2}{dc\lambda^{\alpha+\gamma}} + \frac{\mu(1+\zeta)}{c\lambda^\gamma} + \frac{b^2}{dc\lambda^{\alpha+\gamma-2\beta}} + \frac{1}{d\lambda^\alpha} + \frac{ae^{-\tau\mu(1+\zeta)}}{dc\lambda^{\alpha+\gamma-1}}.$$
$$(7.41)$$

Analogously to the proof of Theorem 7.1 we can conclude the existence of a zero ζ.

Case 2: $2\beta > \alpha + \gamma$.

(7.36) is now equivalent to

$$\omega\left(1 + \frac{\omega^2}{b^2\lambda^{2\beta}} + \frac{d\omega}{b^2\lambda^{2\beta-\alpha}} + \frac{c}{b^2\lambda^{2\beta-\gamma}} + \frac{dc}{b^2\lambda^{2\beta-\alpha-\gamma}} + \frac{ae^{-\tau\omega}}{b^2\lambda^{2\beta-1}}\right)$$
$$= -\frac{ad}{b^2}\lambda^{1+\alpha-2\beta}e^{-\tau\omega}. \qquad (7.42)$$

With the same ansatz (7.38) for ω, where μ solves

$$\mu = -\frac{ad}{b^2}\lambda^{1+\alpha-2\beta}e^{-\tau\mu} \qquad (7.43)$$

we can proceed analogously to Case 1.

Finally, we remark that a delay in (7.9) for θ in the form $bA^\alpha\theta(t - \tau)$ also, and more naturally, leads to ill-posedness, while a delay (only) in coupling terms might lead to well-posedness (cf. [7, 26]).

References

[1] Ammar Khodja, F., Benabdallah, A.: Sufficient conditions for uniform stabilization of second order equations by dynamical controllers. *Dyn. Contin. Discrete Impulsive Syst.* **7** (2000), 207–22.

[2] Ammari, K., Nicaise, S., Pignotti, C.: Feedback boundary stabilization of wave equations with interior delay. *Syst. Control Lett.* **59** (2010), 623–8.

[3] Avalos, G., Lasiecka, S.: Exponential stability of a thermoelastic system without mechanical dissipation. *Rend. Instit. Mat. Univ. Trieste Suppl.* **28** (1997), 1–28.

[4] Denk, R., Racke, R., Shibata, Y.: L_p theory for the linear thermoelastic plate equations in bounded and exterior domains. *Adv. Diff. Eq.* **14** (2009), 685–715.

[5] Denk, R., Racke, R., Shibata, Y.: Local energy decay estimate of solutions to the thermoelastic plate equations in two- and three-dimensional exterior domains. *J. Anal. Appl.* **29** (2010), 21–62.

[6] Dreher, M., Quintanilla, R., Racke, R.: Ill-posed problems in thermomechanics. *Appl. Math. Letters.* **22** (2009), 1374–9.

[7] Fischer, L.: *Instabilität gekoppelter Systeme mit Delay.* Bachelor thesis. University of Konstanz, Germany (2014).

[8] Fridman, E., Nicaise, S., Valein, J.: Stabilization of second order evolution equations with unbounded feedback with time-dependent delay. *SIAM J. Control Optim.* **48** (2010), 5028–52.

[9] Jordan, P.M., Dai, W., Mickens, R.E.: A note on the delayed heat equation: Instability with respect to initial data. *Mech. Res. Comm.* **35** (2008), 414–20.

[10] Khusainov, D., Pokojovy, M., Racke, R.: Strong and mild extra-polated L^2-solutions to the heat equation with constant delay. *SIAM J. Math. Anal.* **47** (1) (2015), 427–54.

[11] Kim, J.U.: On the energy decay of a linear thermoelastic bar and plate. *SIAM J. Math. Anal.* **23** (1992), 889–99.

[12] Kirane, M., Said-Houari, B., Anwar, M.M.: Stability result for the Timoshenko system with a time-varying delay term in the internal feedbacks. *Comm. Pure Appl. Anal.* **10** (2011), 667–86.

[13] Lasiecka, I., Triggiani, R.: Two direct proofs on the analyticity of the S.C. semigroup arising in abstract thermoelastic equations. *Adv. Diff. Eq.* **3** (1998), 387–416.

[14] Lasiecka, I., Triggiani, R.: Analyticity, and lack thereof, of thermo-elastic semigroups. *ESAIM, Proc.* **4** (1998), 199–222.

[15] Lasiecka, I., Triggiani, R.: Analyticity of thermo-elastic semigroups with coupled hinged/Neumann boundary conditions. *Abstract Appl. Anal.* **3** (1998), 153–69.

[16] Lasiecka, I., Triggiani, R.: Analyticity of thermo-elastic semigroups with free boundary conditions. *Annali Scuola Norm. Sup. Pisa* **27** (1998), 457–82.

[17] Liu, Z., Renardy, M.: A note on the equation of a thermoelastic plate. *Appl. Math. Lett.* **8** (1995), 1–6.

[18] Liu, K., Liu, Z.: Exponential stability and analyticity of abstract linear thermoelastic systems. *Z. Angew. Math. Phys.* **48** (1997), 885–904.

[19] Liu, Z., Yong, J.: Qualitative properties of certain C_0 semigroups arising in elastic systems with various dampings. *Adv. Diff. Eq.* **3** (1998), 643–86.

[20] Liu, Z., Zheng, S.: Exponential stability of the Kirchhoff plate with thermal or viscoelastic damping. *Quart. Appl. Math.* **53** (1997), 551–64.

[21] Muñoz Rivera, J.E., Racke, R.: Smoothing properties, decay, and global existence of solutions to nonlinear coupled systems of thermoelastic type. *SIAM J. Math. Anal.* **26** (1995), 1547–63.

[22] Muñoz Rivera, J.E., Racke, R.: Large solutions and smoothing properties for nonlinear thermoelastic systems. *J. Diff. Equat.* **127** (1996), 454–83.

[23] Nicaise, S., Pignotti, C.: Stability and instability results of the wave equation with a delay term in the boundary or internal feedbacks. *SIAM J. Control Optim.* **45** (2006), 1561–85.

[24] Nicaise, S., Pignotti, C.: Stabilization of the wave equation with boundary of internal distributed delay. *Diff. Integr. Equat.* **21** (2008), 935–58.

[25] Prüß, J.: *Evolutionary integral equations and applications.* Monographs Math., **87**, Birkhäuser, Basel (1993).

[26] Racke, R.: Instability in coupled systems with delay. *Comm. Pure Appl. Anal.* **11** (2012), 1753–73.

[27] Russell, D.L.: A general framework for the study of indirect damping mechanisms in elastic systems. *J. Math. Anal. Appl.* **173** (1993), 339–58.

[28] Said-Houari, B., Laskri, Y.: A stability result of a Timoshenko system with a delay term in the internal feedback. *Appl. Math. Comp.* **217** (2010), 2857–69.

DEPARTMENT OF MATHEMATICS AND STATISTICS, UNIVERSITY OF KONSTANZ, 78457 KONSTANZ, GERMANY
E-mail address: lisa.fischer@uni-konstanz.de

DEPARTMENT OF MATHEMATICS AND STATISTICS, UNIVERSITY OF KONSTANZ, 78457 KONSTANZ, GERMANY
E-mail address: reinhard.racke@uni-konstanz.de

8

Controllability of Parabolic Equations by the Flatness Approach

PHILIPPE MARTIN, LIONEL ROSIER,
AND PIERRE ROUCHON

Abstract

We consider linear one-dimensional parabolic equations with space-dependent coefficients that are only measurable and that may be degenerate or singular. We prove the null-controllability with one boundary control by following the flatness approach, which provides explicitly the control and the associated trajectory as series. As an application, we consider the heat equation with a discontinuous coefficient in the principal part. The note ends with a numerical experiment which demonstrates the effectiveness of the method.

1991 Mathematics Subject Classification. 93B05, 93C20

Key words and phrases. Degenerate parabolic equation, singular coefficient, null-controllability, Gevrey functions, flatness

Contents

8.1 Introduction

The null-controllability of parabolic equations has been extensively investigated over several decades. After the pioneering work in [11, 15, 20],
mainly concerned with the one-dimensional case, there has been significant
progress in the general N-dimensional case [13, 14, 19] by using Carleman
estimates. The more recent developments of the theory were concerned
with discontinuous coefficients [1, 2], degenerate coefficients [4, 5, 6, 12],
or singular coefficients [8, 10].

In [1], the authors derived the null-controllability of a linear one-
dimensional parabolic equation with (essentially bounded) measurable
coefficients. The method of proof combined the Lebeau–Robbiano
approach [19] with some complex analytic arguments.

Here, we are concerned with the null-controllability of the system

$$(a(x)u_x)_x + b(x)u_x + c(x)u - \rho(x)u_t = 0, \qquad x \in (0,1),\ t \in (0,T), \quad (8.1)$$
$$\alpha_0 u(0,t) + \beta_0 (au_x)(0,t) = 0, \qquad t \in (0,T), \quad\quad\quad (8.2)$$
$$\alpha_1 u(1,t) + \beta_1 (au_x)(1,t) = h(t), \qquad t \in (0,T), \quad\quad\quad (8.3)$$
$$u(x,0) = u_0(x), \qquad x \in (0,1), \quad\quad\quad (8.4)$$

where $(\alpha_0, \beta_0), (\alpha_1, \beta_1) \in \mathbb{R}^2 \setminus \{(0,0)\}$ are given, $u_0 \in L^2(0,1)$ is the initial
state, and $h \in L^2(0,T)$ is the control input.

The given functions a, b, c, ρ will be assumed to fulfill the following
conditions

$$a(x) > 0 \quad \text{and} \quad \rho(x) > 0 \quad \text{for a.e. } x \in (0,1), \quad\quad (8.5)$$
$$\left(\frac{1}{a}, \frac{b}{a}, c, \rho\right) \in [L^1(0,1)]^4, \quad\quad\quad (8.6)$$
$$\exists K \geq 0, \quad \frac{c(x)}{\rho(x)} \leq K \quad \text{for a.e. } x \in (0,1), \quad\quad (8.7)$$
$$\exists p \in (1, \infty], \quad a^{1-\frac{1}{p}}\rho \in L^p(0,1). \quad\quad\quad (8.8)$$

The assumptions (8.5)–(8.8) are clearly less restrictive than the assumptions
from [1]:

$$a, b, c, \rho \in L^\infty(0,1) \quad \text{and} \quad a(x) > \varepsilon,\ \rho(x) > \varepsilon > 0 \quad \text{for a.e. } x \in (0,1) \quad (8.9)$$

for some $\varepsilon > 0$.

Let us introduce some notations. Let B be a Banach space with norm
$\|\cdot\|_B$. For any $t_1 < t_2$ and $s \geq 0$, we denote by $G^s([t_1, t_2], B)$ the class of

(Gevrey) functions $u \in C^\infty([t_1, t_2], B)$ for which there exist positive constants M, R such that

$$\|u^{(p)}(t)\|_B \leq M \frac{p!^s}{R^p} \quad \forall t \in [t_1, t_2], \ \forall p \geq 0. \tag{8.10}$$

When $(B, \|\cdot\|_B) = (\mathbb{R}, |\cdot|)$, $G^s([t_1, t_2], B)$ is merely denoted $G^s([t_1, t_2])$. Let

$$L^1_\rho := \{u : (0,1) \to \mathbb{R}; \|u\|_{L^1_\rho} := \int_0^1 |u(x)|\rho(x)dx < \infty\}.$$

Note that $L^2(0,1) \subset L^1_\rho$ if $\rho \in L^2(0,1)$. The main result in this note is the following:

Theorem 8.1 *Let the functions* $a, b, c, \rho : (0,1) \to \mathbb{R}$ *satisfy* (8.5)–(8.8) *for some numbers* $K \geq 0$, $p \in (1, \infty]$. *Let* $(\alpha_0, \beta_0), (\alpha_1, \beta_1) \in \mathbb{R}^2 \setminus \{(0,0)\}$ *and* $T > 0$. *Pick any* $u_0 \in L^1_\rho$ *and any* $s \in (1, 2 - 1/p)$. *Then there exists a function* $h \in G^s([0,T])$, *that may be given explicitly as a series, such that the solution* u *of* (8.1)–(8.4) *satisfies* $u(\cdot, T) = 0$. *Moreover* $u, au_x \in G^s([\varepsilon, T], W^{1,1}(0,1))$ *for all* $\varepsilon \in (0,T)$.

Clearly, Theorem 8.1 can be applied to parabolic equations with discontinuous coefficients that may be degenerate or singular at a point (or more generally at a sequence of points). The proof of it is not based on the classical duality approach, in the sense that it does not rely on the proof of some observability inequality for the adjoint equation. It follows the flatness approach developed in [16, 17, 18, 21, 22, 24, 25, 27, 28, 29]. This direct approach gives explicitly both the control and the trajectory as series, which leads to efficient numerical schemes by taking partial sums in the series [24]. The flatness approach was used in [28] to improve the results in [11] concerning the reachable states for the heat equation. It was also successfully applied to the Schrödinger equation in [23]. Let us describe its main steps. In the first step, following [1], we show that after a series of changes of dependent/independent variables, system (8.1)–(8.4) can be put into the canonical form

$$u_{xx} - \rho(x)u_t = 0, \quad x \in (0,1), \ t \in (0,T), \tag{8.11}$$

$$\alpha_0 u(0,t) + \beta_0 u_x(0,t) = 0, \quad t \in (0,T), \tag{8.12}$$

$$\alpha_1 u(1,t) + \beta_1 u_x(1,t) = h(t), \quad t \in (0,T), \tag{8.13}$$

$$u(x,0) = u_0(x), \quad x \in (0,1), \tag{8.14}$$

where $\rho(x) > 0$ a.e. in $(0, 1)$ and $\rho \in L^p(0, 1)$ with $p \in (1, \infty]$. In the second step, following [21, 24], we seek u in the form

$$u(x, t) = \sum_{n \geq 0} e^{-\lambda_n t} e_n(x), \quad x \in (0, 1), \ t \in [0, \tau], \tag{8.15}$$

$$u(x, t) = \sum_{i \geq 0} y^{(i)}(t) g_i(x), \quad x \in (0, 1), \ t \in [\tau, T], \tag{8.16}$$

where $\tau \in (0, T)$ is any intermediate time; $e_n : (0, 1) \to \mathbb{R}$ (resp. $\lambda_n \in \mathbb{R}$) denotes the nth *eigenfunction* (resp. *eigenvalue*) associated with (8.11)–(8.13) and satisfying [16, 17]

$$-e_n'' = \lambda_n \rho e_n, \quad x \in (0, 1) \tag{8.17}$$

$$\alpha_0 e_n(0) + \beta_0 e_n'(0) = 0, \tag{8.18}$$

$$\alpha_1 e_n(1) + \beta_1 e_n'(1) = 0, \tag{8.19}$$

while $g_i : (0, 1) \to \mathbb{R}$ is defined inductively as the solution to the Cauchy problem

$$g_0'' = 0, \quad x \in (0, 1) \tag{8.20}$$

$$\alpha_0 g_0(0) + \beta_0 g_0'(0) = 0, \tag{8.21}$$

$$\beta_0 g_0(0) - \alpha_0 g_0'(0) = 1 \tag{8.22}$$

for $i = 0$, and to the Cauchy problem

$$g_i'' = \rho g_{i-1}, \quad x \in (0, 1), \tag{8.23}$$

$$g_i(0) = 0, \tag{8.24}$$

$$g_i'(0) = 0 \tag{8.25}$$

for $i \geq 1$. Expanding u on generating functions as in (8.16) rather than on powers of x as in [18, 21] was introduced in [17] and studied in [16].

The fact that the generating function g_i is defined as the solution of a *Cauchy problem*, rather than the solution of a *boundary-value problem*, is crucial in the analysis developed here. First, it allows to prove that *every* initial state in the space L_ρ^1 (and not only states in some restricted class of Gevrey functions) can be driven to 0 in time T. Secondly, from (8.23)–(8.25), we see by an easy induction on i that for $\rho \in L^\infty(0, 1)$, the function g_i is uniformly bounded by $C/(2i)!$, and hence the series in (8.16) is indeed convergent when $y \in G^s([\tau, T])$ with $1 < s < 2$.

The corresponding control function h is given explicitly as

$$h(t) = \begin{cases} 0 & \text{if } 0 \leq t \leq \tau, \\ \sum_{i \geq 0} y^{(i)}(t)(\alpha_1 g_i(1) + \beta_1 g_i'(1)) & \text{if } \tau < t \leq T. \end{cases}$$

It is easy to see that the function $u(x, t)$ defined in (8.16) satisfies (formally) (8.11), and also the condition $u(x, T) = 0$ if $y^{(i)}(T) = 0$ for all $i \in \mathbb{N}$, so that the null-controllability can be established for *some* initial states. The main issue is then to extend it to *every* initial state $u_0 \in L_\rho^1$. Following [21, 22, 24], we first use the strong smoothing effect of the heat equation to smooth out the state function in the time interval $(0, \tau)$. Next, to ensure that the two expressions of u given in (8.15)–(8.16) coincide at $t = \tau$, we have to relate the eigenfunctions e_n to the generating functions g_i.

It will be shown that any eigenfunction e_n can be expanded in terms of the generating functions g_i as

$$e_n(x) = \zeta_n \sum_{i \geq 0} (-\lambda_n)^i g_i(x) \tag{8.26}$$

with $\zeta_n \in \mathbb{R}$. Note that, for $\rho \equiv 1$ and $(\alpha_0, \beta_0, \alpha_1, \beta_1) = (0, 1, 0, 1)$, $\lambda_n = (n\pi)^2$ for all $n \geq 0$, $e_0(x) = 1$ and $e_n(x) = \sqrt{2} \cos(n\pi x)$ for $n \geq 0$ while $g_i(x) = x^{2i}/(2i)!$, so that (8.26) for $n \geq 1$ is nothing but the classical Taylor series expansion of $\cos(n\pi x)$ at $x = 0$:

$$\cos(n\pi x) = \sum_{i \geq 0} (-1)^i \frac{(n\pi x)^{2i}}{(2i)!}. \tag{8.27}$$

Thus (8.26) can be seen as a natural extension of (8.27), in which the generating functions g_i, *a priori* not smoother than $W^{2,p}(0, 1)$, replace the functions $x^{2i}/(2i)!$.

The condition (8.8) is used to prove the estimate

$$|g_i(x)| \leq \frac{C}{R^{2i}(i!)^{2 - \frac{1}{p}}}$$

needed to ensure the convergence of the series in (8.16) when $y \in G^s([\tau, T])$ with $1 < s < 2 - 1/p$.

Theorem 8.1 applies in particular to any system

$$(a(x)u_x)_x - u_t = 0, \qquad x \in (0, 1), \ t \in (0, T), \tag{8.28}$$
$$\alpha_0 u(0, t) + \beta_0 (au_x)(0, t) = 0, \qquad t \in (0, T), \tag{8.29}$$
$$\alpha_1 u(1, t) + \beta_1 (au_x)(1, t) = h(t), \qquad t \in (0, T), \tag{8.30}$$
$$u(x, 0) = u_0(x), \qquad x \in (0, 1), \tag{8.31}$$

where $a(x) > 0$ for a.e. $x \in (0, 1)$ and $a + 1/a \in L^1(0, 1)$. This includes the case where a is measurable, positive, and essentially bounded together with its inverse (but not necessarily piecewise continuous), and the case where $a(x) = x^r$ with $-1 < r < 1$. (Actually any $r \leq -1$ is also admissible, by picking $p > 1$ sufficiently close to 1 in (8.8).) Note that our result applies as well to $a(x) = (1 - x)^r$ with $0 < r < 1$, yielding a positive null-controllability result when the control is applied at the point $(x = 1)$ where the diffusion coefficient degenerates (see [5, section 2.7]). Note also that the coefficient $a(x)$ is allowed to be degenerate/singular at a *sequence* of points: consider, e.g., $a(x) := |\sin(x^{-1})|^r$ with $-1 < r < 1$. Then $a + 1/a \in L^1(0, 1)$.

The null-controllability of (8.28)–(8.31) for $a(x) = x^r$ with $0 < r < 2$ was established (in appropriate spaces) in [5].

Another important family of heat equations with variable coefficients is those with inverse square potential localized at the boundary, namely

$$u_{xx} + \frac{\mu}{x^2}u - u_t = 0, \qquad x \in (0, 1), \ t \in (0, T), \qquad (8.32)$$

$$u(0, t) = 0, \qquad t \in (0, T), \qquad (8.33)$$

$$\alpha_1 u(1, t) + \beta_1 u_x(1, t) = h(t), \qquad t \in (0, T), \qquad (8.34)$$

$$u(x, 0) = u_0(x), \qquad x \in (0, 1), \qquad (8.35)$$

where $\mu \in \mathbb{R}$ is a given number. Note that Theorem 8.1 cannot be applied to (8.32)–(8.35), for $c(x) = \mu x^{-2}$ is not integrable on $(0, 1)$. It was proved in [8] that (8.32)–(8.35) is null-controllable in $L^2(0, 1)$ when $\mu \leq 1/4$ by combining Carleman inequalities to Hardy inequalities. We note that this result can be retrieved by the flatness approach as well.

Theorem 8.2 *Let $\mu \in (0, 1/4]$, $(\alpha_1, \beta_1) \in \mathbb{R}^2 \setminus \{(0, 0)\}$, $T > 0$, and $\tau \in (0, T)$. Pick any $u_0 \in L^2(0, 1)$ and any $s \in (1, 2)$. Then there exists a function $h \in G^s([0, T])$ with $h(t) = 0$ for $0 \leq t \leq \tau$ and such that the solution u of (8.32)–(8.35) satisfies $u(T, \cdot) = 0$. Moreover, $u \in G^s([\varepsilon, T], W^{1,1}(0, 1))$ for all $\varepsilon \in (0, T)$. Finally, if $0 \leq \mu < 1/4$ and $r > (1 + \sqrt{1 - 4\mu})/2$, then $x^r u_x \in G^s([\varepsilon, T], W^{1,1}(0, 1))$ for all $\varepsilon \in (0, T)$.*

The note is organized as follows. In Section 8.2, we give a sketch of the proof of Theorem 8.1. Section 8.3 is concerned with the numerical control of a heat equation with discontinuous coefficients, that may serve as a model for the heat conduction of a one-dimensional rod with constant thermal properties.

8.2 Sketch of the Proof of Theorem 8.1

8.2.1 Reduction to the Canonical Form (8.11)–(8.14)

Let a, b, c, ρ, and p be as in (8.5)–(8.8). Set

$$B(x) := \int_0^x \frac{b(s)}{a(s)} ds,$$

$$\tilde{a}(x) := a(x)e^{B(x)}$$

$$\tilde{c}(x) := (K\rho(x) - c(x))e^{B(x)}.$$

Then $B \in W^{1,1}(0,1)$, $\tilde{c} \in L^1(0,1)$, and

$$\tilde{a}(x) > 0 \quad \text{and} \quad \tilde{c}(x) \geq 0 \quad \text{for a.e. } x \in (0,1).$$

We introduce the solution v to the elliptic boundary value problem

$$-(\tilde{a}v_x)_x + \tilde{c}v = 0, \quad x \in (0,1), \tag{8.36}$$

$$v(0) = v(1) = 1, \tag{8.37}$$

and set

$$u_1(x, t) := e^{-Kt}u(x, t), \quad u_2(x, t) := \frac{u_1(x, t)}{v(x)}.$$

Finally, let

$$L := \int_0^1 (a(s)v^2(s)e^{B(s)})^{-1} ds, \quad y(x) := \frac{1}{L}\int_0^x (a(s)v^2(s)e^{B(s)})^{-1} ds \tag{8.38}$$

and

$$\hat{u}(y, t) := u_2(x, t), \quad \hat{\rho}(y) := L^2 a(x)v^4(x)e^{2B(x)}\rho(x) \tag{8.39}$$

for $0 < t < 1$, $y = y(x)$ with $0 < x < 1$. Then the following result holds.

Proposition 8.3
(i) $v \in W^{1,1}(0,1)$ and $0 < v(x) \leq 1$ for all $x \in [0,1]$;
(ii) $y: [0,1] \to [0,1]$ is an increasing bijection with $y, y^{-1} \in W^{1,1}(0,1)$;
(iii) $\hat{\rho}(y) > 0$ for a.e. $y \in (0,1)$, and $\hat{\rho} \in L^p(0,1)$;
(iv) $\hat{u}$ solves the system

$$\hat{u}_{yy} - \hat{\rho}\hat{u}_t = 0, \quad y \in (0,1), \ t \in (0,T), \tag{8.40}$$

$$\hat{\alpha}_0\hat{u}(0, t) + \hat{\beta}_0\hat{u}_y(0, t) = 0, \quad t \in (0,T), \tag{8.41}$$

$$\hat{\alpha}_1\hat{u}(1, t) + \hat{\beta}_1\hat{u}_y(1, t) = \hat{h}(t) := e^{-Kt}h(t), \quad t \in (0,T), \tag{8.42}$$

$$\hat{u}(y(x), 0) = \frac{u_0(x)}{v(x)}, \quad x \in (0,1), \tag{8.43}$$

for some $(\hat{\alpha}_0, \hat{\beta}_0), (\hat{\alpha}_1, \hat{\beta}_1) \in \mathbb{R}^2 \setminus \{(0,0)\}$.

Proof (i) Let $l:=\int_0^1 ds/\tilde{a}(s)$ and $z(x):=l^{-1}\int_0^x ds/\tilde{a}(s)$. Then, one can see that z is an increasing continuous bijection from $[0,1]$ to $[0,1]$, and (using a result due to Zareckii, see, e.g., [3]) that $z^{-1} \in W^{1,1}(0,1)$. Introduce $w(z):= v(x(z))$, which solves the boundary value problem

$$-\frac{d^2w}{dz^2} + (l^2\tilde{a}\tilde{c})(x(z))w = 0, \quad z \in (0,1),$$
$$w(0) = w(1) = 1.$$

It can be seen that $0 < w \le 1$ on $[0,1]$. The other properties (ii), (iii), and (iv) follow by direct calculations (see [27] for more details).

8.2.2 Null-Controllability of the Control Problem (8.11)–(8.14)

Assume given $p \in (1,\infty]$, $\rho \in L^p(0,1)$ with $\rho(x) > 0$ for a.e. $x \in (0,1)$, and $(\alpha_0,\beta_0),(\alpha_1,\beta_1) \in \mathbb{R}^2 \setminus \{(0,0)\}$. Let $' = d/dx$, and let

$$L_\rho^2 := \{f: (0,1) \to \mathbb{R}; \|f\|_{L_\rho^2}^2 := \int_0^1 |f(x)|^2\rho(x)dx < \infty\}.$$

Then the following result holds.

Proposition 8.4 Let $p,\rho,\alpha_0,\beta_0,\alpha_1$, and β_1 be as above. Then there are a sequence $(e_n)_{n\ge0}$ in L_ρ^2 and a sequence $(\lambda_n)_{n\ge0}$ in $\mathbb{R}$ such that

(i) $(e_n)_{n\ge0}$ is an orthonormal basis in L_ρ^2;
(ii) For all $n \ge 0$, $e_n \in W^{2,p}(0,1)$ and e_n solves

$$-e_n'' = \lambda_n\rho e_n \quad \text{in } (0,1), \tag{8.44}$$
$$\alpha_0 e_n(0) + \beta_0 e_n'(0) = 0, \tag{8.45}$$
$$\alpha_1 e_n(1) + \beta_1 e_n'(1) = 0. \tag{8.46}$$

(iii) The sequence $(\lambda_n)_{n\ge0}$ is strictly increasing, and for some constant $C > 0$

$$\lambda_n \ge Cn \quad \text{for } n \gg 1. \tag{8.47}$$

Proof Introduce for $\lambda^* \gg 1$ the boundary-value problem

$$-u'' + \lambda^*\rho u = \rho f \quad \text{in } (0,1), \tag{8.48}$$
$$\alpha_0 u(0) + \beta_0 u'(0) = 0, \tag{8.49}$$
$$\alpha_1 u(1) + \beta_1 u'(1) = 0. \tag{8.50}$$

Then, introducing a variational formulation and applying Lax–Milgram theorem, we obtain the existence and uniqueness of a solution $u \in W^{2,1}(0,1)$ of (8.48)–(8.50) for any $f \in L_\rho^2$. The results in (i)–(ii) follow from an application of the spectral theorem. Finally, (iii) is established by using Prüfer substitution. We refer the reader to [27] for more details.

We now turn our attention to the generating functions g_i $(i \geq 0)$ defined along (8.20)–(8.25).

Proposition 8.5

(i) $g_0(x) = (\alpha_0^2 + \beta_0^2)^{-1}(\beta_0 - \alpha_0 x)$.

(ii) There are some constants $C, R > 0$ such that

$$\|g_i\|_{W^{2,p}(0,1)} \leq \frac{C}{R^i (i!)^{2 - \frac{1}{p}}} \qquad \forall i \geq 0. \tag{8.51}$$

Proof (i) is obvious and (ii) is obtained inductively from the formula

$$g_i(x) = \int_0^x \left(\int_0^s \rho(\sigma) g_{i-1}(\sigma) d\sigma \right) ds$$

(see [27] for more details).

The fact that we can expand the eigenfunctions in terms of the generating functions is detailed in the following

Proposition 8.6 There is some sequence $(\zeta_n)_{n \geq 0}$ of real numbers such that for all $n \geq 0$

$$e_n = \zeta_n \sum_{i \geq 0} (-\lambda_n)^i g_i \quad \text{in } W^{2,p}(0,1). \tag{8.52}$$

Furthermore, for some constant $C > 0$, we have

$$|\zeta_n| \leq C(1 + |\lambda_n|^{\frac{3}{2}}) \quad \forall n \geq 0. \tag{8.53}$$

Proof Let $\tilde{e} := \zeta_n \sum_{i \geq 0} (-\lambda_n)^i g_i$, where $\zeta_n \in \mathbb{R}$. Then we infer from (8.20)–(8.25) that $\tilde{e}'' = -\lambda_n \rho \tilde{e}$ and that

$$\alpha_0 \tilde{e}(0) + \beta_0 \tilde{e}'(0) = 0.$$

If we pick $\zeta_n := \beta_0 e_n(0) - \alpha_0 e_n'(0)$, then we easily see that $e_n = \tilde{e}$. The estimate (8.53) is proved by direct calculations. We refer the reader to [27] for more details.

Since $p > 1$, for any $s \in (1, 2 - \frac{1}{p})$ and any $0 < \tau < T$, one may pick a function $\varphi \in G^s([0, 2T])$ such that

$$\varphi(t) = \begin{cases} 1 & \text{if } t \leq \tau, \\ 0 & \text{if } t \geq T. \end{cases}$$

We are in a position to derive the null-controllability of (8.11)–(8.14). Let $u_0 \in L_\rho^2$. Since $(e_n)_{n \geq 0}$ is an orthonormal basis in L_ρ^2, we can write

$$u_0 = \sum_{n \geq 0} c_n e_n \quad \text{in } L_\rho^2 \tag{8.54}$$

with $\sum_{n \geq 0} |c_n|^2 < \infty$. Let

$$y(t) := \varphi(t) \sum_{n \geq 0} c_n \zeta_n e^{-\lambda_n t} \quad \text{for } t \in [\tau, T] \tag{8.55}$$

and

$$u(x, t) = \begin{cases} \sum_{n \geq 0} c_n e^{-\lambda_n t} e_n(x) & \text{if } 0 \leq t \leq \tau, \\ \sum_{i \geq 0} y^{(i)}(t) g_i(x) & \text{if } \tau < t \leq T. \end{cases} \tag{8.56}$$

The main result in this section is the following:

Theorem 8.7 *Let $p \in (1, \infty]$, $\rho \in L^p(0, 1)$ with $\rho(x) > 0$ for a.e. $x \in (0, 1)$, $T > 0$, $\tau \in (0, T)$, and $(\alpha_0, \beta_0), (\alpha_1, \beta_1) \in \mathbb{R}^2 \setminus \{(0, 0)\}$. Let $u_0 \in L_\rho^2$ be decomposed as in (8.54), and let y be as in (8.55). Then $y \in G^s([\tau, T])$, and the control*

$$h(t) = \begin{cases} 0 & \text{if } 0 \leq t \leq \tau, \\ \sum_{i \geq 0} y^{(i)}(t)(\alpha_1 g_i(1) + \beta_1 g_i'(1)) & \text{if } \tau < t \leq T \end{cases} \tag{8.57}$$

is such that the solution u of (8.11)–(8.14) satisfies $u(\cdot, T) = 0$. Moreover, u is given by (8.56), $h \in G^s([0, T])$, and $u \in C([0, T], L_\rho^2) \cap G^s([\varepsilon, T], W^{2,p}(0, 1))$ for all $0 < \varepsilon \leq T$.

Proof It is clear that the function u solves formally (8.11)–(8.14) on $(0, \tau)$ and on (τ, T), together with $u(\cdot, T) = 0$. The main concern is thus the convergence of the two series in (8.56) and the fact that they coincide at $t = \tau$. See [27] for the details.

Theorem 8.1 follows from Theorem 8.7. Modifying slightly the first step in the proof of Theorem 8.1, we can reduce (8.32)–(8.35) to the canonical form (8.11)–(8.14), so that the conclusion of Theorem 8.2 follows from Theorem 8.7.

As a possible application, we consider the boundary control by the flatness approach of radial solutions of the heat equation in the ball $B(0, 1) \subset \mathbb{R}^N$ ($2 \leq N \leq 3$). Using the radial coordinate $r = |x|$, we thus consider the system

$$u_{rr} + \frac{N-1}{r}u_r - u_t = 0, \quad r \in (0,1), \ t \in (0,T), \tag{8.58}$$

$$u_r(0,t) = 0, \quad t \in (0,T), \tag{8.59}$$

$$\alpha_1 u(1,t) + \beta_1 u_r(1,t) = h(t), \quad t \in (0,T) \tag{8.60}$$

$$u(r,0) = u_0(r), \quad r \in (0,1). \tag{8.61}$$

Note that Theorem 8.1 cannot be applied directly to (8.58)–(8.61), for (8.7) fails. (Note that, in sharp contrast, the control on a ring-shaped domain $\{r_0 < |x| < r_1\}$ with $r_1 > r_0 > 0$ is fully covered by Theorem 8.1, the coefficients in (8.58) being then smooth and bounded.)

We use the following change of variables from [9] which allows to remove the term with the first order derivative in r in (8.58):

$$u(r,t) = \tilde{u}(r,t)\exp\left(-\frac{1}{2}\int_0^r \frac{N-1}{s}ds\right) = \frac{\tilde{u}(r,t)}{r^{\frac{N-1}{2}}}. \tag{8.62}$$

Then (8.58) becomes

$$\tilde{u}_{rr} + \frac{(N-1)(3-N)}{4}\frac{\tilde{u}}{r^2} - \tilde{u}_t = 0. \tag{8.63}$$

This equation has to be supplemented with the boundary/initial conditions

$$\tilde{u}(0,t) = 0, \quad t \in (0,T), \tag{8.64}$$

$$(\alpha_1 - \frac{N-1}{2}\beta_1)\tilde{u}(1,t) + \beta_1\tilde{u}_r(1,t) = h(t), \quad t \in (0,T), \tag{8.65}$$

$$\tilde{u}(r,0) = r^{\frac{N-1}{2}}u_0(r), \quad r \in (0,R). \tag{8.66}$$

For $N = 3$, (8.63) reduces to the simple heat equation $\tilde{u}_{rr} - \tilde{u}_t = 0$ to which Theorem 8.1 can be applied.

For $N = 2$, (8.63)–(8.66) is of the form (8.32)–(8.35) with $\mu = 1/4$. Therefore Theorem 8.2 can be applied.

8.3 Numerical Control of the 1D Heat Equation with Discontinuous Coefficients

We consider the heat conduction in a one-dimensional rod made of two sections with constant thermal properties. Without restriction, we can assume the rod has length 1, with one section of length X and the other of length $1 - X$. The evolution of the temperature u is given by the heat equation

$$\rho\theta_t(x,t) = (a\theta_x)_x(x,t).$$

a and ρ are the piecewise constant functions on $(0, 1)$:

$$(a(x), \rho(x)) := \begin{cases} (a_0, \rho_0), & 0 < x < X, \\ (a_1, \rho_1), & X < x < 1 \end{cases}$$

where a_0, a_1, ρ_0 and ρ_1 are strictly positive constants.

At the 0-end, the rod is submitted to the constant ambient temperature θ_0, and at the 1-end to a time-varying heat source (the control input) of temperature $\theta_1(t)$. The heat flux $-a\theta_x$ at the ends obeys the convection conditions

$$-(a\theta_x)(0, t) = h_0(\theta_0 - \theta(0, t)),$$
$$-(a\theta_x)(1, t) = h_1(\theta(1, t) - \theta_1(t))$$

with h_0, h_1 some positive constants. Setting $u(x, t) := \theta(x, t) - \theta_0$ and taking as input $h(t) := \theta_1(t) - \theta_0$ results in the boundary value problem

$$(a(x)u_x)_x - \rho(x)u_t = 0, \qquad x \in (0, 1), \; t \in (0, T), \qquad (8.67)$$
$$\alpha_0 u(0, t) + \beta_0(au_x)(0, t) = 0, \qquad t \in (0, T), \qquad (8.68)$$
$$\alpha_1 u(1, t) + \beta_1(au_x)(1, t) = h(t), \qquad t \in (0, T), \qquad (8.69)$$
$$u(x, 0) = u_0(x), \qquad x \in (0, 1), \qquad (8.70)$$

where the constants $\alpha_0, \beta_0, \alpha_1, \beta_1$ satisfy $\alpha_0^2 + \beta_0^2 > 0$, $\alpha_1^2 + \beta_1^2 > 0$, $\alpha_0\beta_0 \le 0$ and $\alpha_1\beta_1 \ge 0$. Notice the two limiting cases: $\beta_i = 0$ (Dirichlet conditions), obtained when taking $h_i \to \infty$; $\alpha_i = 0$ (Neumann conditions), obtained when taking as control input $h(t) := h_1(\theta_1(t) - \theta(1, t))$ and letting $h_0 = 0$.

Included in the formulation of the system is the fact that a solution u and its quasi-derivative au_x are continuous on $[0, 1]$, and in particular at $x = X$ (whereas u_x will in general be discontinuous at X). We could thus rewrite (8.67) more explicitly as the piecewise constant heat equation

$$\begin{cases} u_t(x, t) = \dfrac{a_0}{\rho_0} u_{xx}(x, t), & 0 < x < X, \\ u_t(x, t) = \dfrac{a_1}{\rho_1} u_{xx}(x, t), & X < x < 1 \end{cases}$$

together with the so-called interface conditions

$$u(X^-, t) = u(X^+, t),$$
$$a_0 u_x(X^-, t) = a_1 u_x(X^+, t).$$

From the flatness approach developed in the previous sections, we know that the trajectory u of (8.67)–(8.70) is given explicitly in (8.15) and (8.16), the eigenfunctions e_n's being as in (8.17)–(8.19), and the generating functions g_i's being as in (8.20)–(8.25).

We refer the reader to [25] for the details concerning the numerical computation of the e_n's and the g_i's.

The control input in the time interval (τ, T) reads

$$h(t) := \sum_{i \geq 0} (\alpha_1 g_i(1) + \beta_1 g_i'(1)) y^{(i)}(t) \qquad (8.71)$$

where

$$y(t) := \phi_s \left(\frac{t - \tau}{T - \tau} \right) \sum_{n \geq 0} c_n \zeta_n e^{-\lambda_n t} \qquad (8.72)$$

and the c_n's denote the Fourier coefficients of the initial state decomposed along the e_n's:

$$u_0(x) := \sum_{n \geq 0} c_n e_n(x).$$

The function ϕ_s in (8.72) is the Gevrey "step function"

$$\phi_s(t) := \begin{cases} 1 & \text{if } t \leq 0, \\ 0 & \text{if } t \geq 1, \\ 1 - \dfrac{\int_0^t \varphi_s(z) dz}{\int_0^1 \varphi_s(z) dz} & \text{if } t \in (0, 1) \end{cases}$$

where the "bump function" φ_s is defined as

$$\varphi_s(t) = \begin{cases} 0 & \text{if } t \notin (0, 1), \\ \exp \left(\dfrac{-1}{M t^k (1 - t)^k} \right) & \text{if } t \in (0, 1) \end{cases}$$

with $k = (s - 1)^{-1} > 1$ and $M > 0$ some constants. It is well known that both φ_s and ϕ_s are Gevrey of order s, with $s \in (1, 2)$.

A practical problem when implementing the control (8.72) is to evaluate sufficiently many derivatives of ϕ_s, i.e., of φ_s. Numerical computations (with, e.g., finite differences) or symbolic computations cannot be used in practice to evaluate more than 20 derivatives. Nevertheless, many more derivatives can be computed with accuracy by proceeding *inductively*. We first note that

$$p^{k+1} \dot{\varphi}_s = k \dot{p} \varphi_s \qquad (8.73)$$

where $p(t) := M^{\frac{1}{k}} t(1 - t)$ is a polynomial function of degree 2 (hence its derivatives of order > 2 are zero).

Derivating i times in (8.73) and using Leibniz's rule results in

$$p^{k+1} \varphi_s^{(i+1)} + \sum_{j=1}^{i} \binom{i}{j} (p^{k+1})^{(j)} \varphi_s^{(i+1-j)} = k(\dot{p} \varphi_s^{(i)} + i \ddot{p} \varphi_s^{(i-1)}).$$

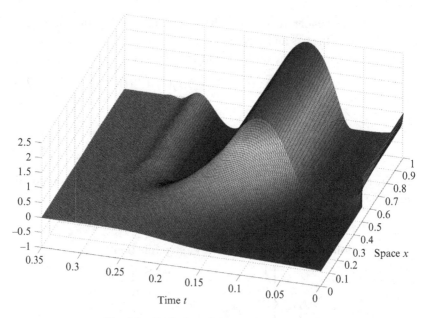

Fig. 8.1.   Evolution of the temperature $u(x, t)$

This formula gives $\varphi_s^{(i+1)}$ in terms of $\varphi_s^{(0)}, \ldots, \varphi_s^{(i)}$, and the derivatives of $P := p^{k+1}$ that can be obtained similarly, by applying Leibniz' rule to both sides of

$$p\dot{P} = (k + 1)\dot{p}P.$$

Note that, in order to avoid computing ratios of very large numbers, it is better in practice to use recursion formulae for $\tilde{\varphi}_s^{(i)} := \varphi_s^{(i)}/(2i)!$ and for $\tilde{P}^{(i)} := P^{(i)}/i!$ (see [25]).

Using this procedure, about 140 derivatives can be efficiently determined with Matlab® double-precision arithmetics.

We conclude this note with some numerical simulation. We use as parameters: $X = 1/2$, $a_0 = 10/19$, $\rho_0 = 15/8$, $a_1 = 10$, $\rho_1 = 1/8$, $\alpha_0 = \cos(\pi/3)$, $\beta_0 = -\sin(\pi/3)$, $\alpha_1 = \cos(\pi/4)$, $\beta_1 = \sin(\pi/4)$. We take as initial condition $u_0(x) := \frac{1}{2}1_{(1/2,1)}(x) - \frac{1}{2}1_{(0,1/2)}(x)$, and as control times: $\tau = 0.05$ and $T = 0.35$. We pick also $s = 1.65$ and $M = 2$. The series for h and y in (8.71) and (8.72) were truncated at a "large enough" order for a good accuracy, namely $\bar{\imath} = 130$ and $\bar{n} = 60$; a fairly large $\bar{\imath}$ is needed here because $(T - \tau)a_0/\rho_0$ is rather small. Figure 8.1 shows the resulting temperature u; the discontinuity of u_x at $x = X$ is clearly visible. We refer the interested

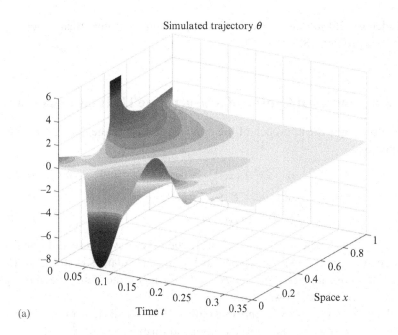

(a)

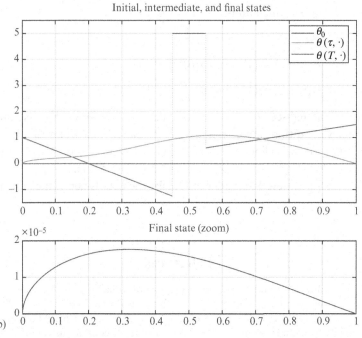

(b)

Fig. 8.2. (a) Evolution of θ ($\gamma = 1/2$) and (b) corresponding initial, intermediate, and final states

reader to [25] for more details about the numerical investigation of the control problem (8.67)–(8.70).

8.4 Numerical Control of a Degenerate 1D Heat Equation

We now apply the approach of the previous section to the degenerate heat equation

$$\theta_t(x,t) - \left(x^\gamma \theta_x(x,t)\right)_x = 0, \quad (x,t) \in (0,1) \times (0,T) \quad (8.74)$$

$$\alpha_0 \theta(0,t) + \beta_0 \left(x^\gamma \theta_x(x,t)\right)\big|_{x=0} = u(t), \quad (8.75)$$

$$\alpha_1 \theta(1,t) + \beta_1 \theta_x(1,t) = 0, \quad (8.76)$$

with $0 \le \gamma < 1$. The controllability properties of this system (with Dirichlet boundary conditions) are studied in [7]; in particular, it is shown in [7] that the cost in terms of the $H^1(0,T)$-norm of the control achieving null-controllability blows up as γ tends to 1^-.

The assumptions (8.7) and (8.8) are clearly satisfied, with $K=0$ and $p=\infty$. The system is put into its canonical form by setting $\tilde{x} := x^{1-\gamma}$ and $\tilde{\theta}(\tilde{x},t) := \theta(x,t)$, which immediately yields

$$(1-\gamma)^{-2} \tilde{x}^{\frac{\gamma}{1-\gamma}} \tilde{\theta}_t(\tilde{x},t) - \tilde{\theta}_{\tilde{x}\tilde{x}}(\tilde{x},t) = 0, \quad (\tilde{x},t) \in (0,1) \times (0,T)$$

$$\alpha_0 \tilde{\theta}(0,t) + (1-\gamma)\beta_0 \tilde{\theta}_{\tilde{x}}(0,t) = u(t),$$

$$\alpha_1 \tilde{\theta}(1,t) + (1-\gamma)\beta_1 \tilde{\theta}_{\tilde{x}}(1,t) = 0.$$

For the actual scenario ($\gamma=0.5$, $T=0.35$, $\tau = T/10$, $s=1.7$), Neumann boundary conditions are used ($\alpha_0 = \alpha_1 = 0$, $\beta_0 = \beta_1 = 1$). The initial condition is the piecewise linear function $\theta_0(x) := 1 - 5x$, $x \in (0,0.45)$, $\theta_0(x) := 5.5$, $x \in (0.45,0.55)$, and $\theta_0(x) := 2x - 0.5$, $x \in (0.55,1)$.

We refer the interested reader to [26] for the details about the numerical scheme.

References

[1] G. Alessandrini and L. Escauriaza. Null-controllability of one-dimensional parabolic equations. *ESAIM Control Optim. Calc. Var.*, 14(2):284–93, 2008.

[2] A. Benabdallah, Y. Dermenjian, and J. Le Rousseau. Carleman estimates for the one-dimensional heat equation with a discontinuous coefficient and applications to controllability and an inverse problem. *J. Math. Anal. Appl.*, 336(2):865–7, 2007.

[3] V. I. Bogachev. *Measure Theory*. Springer-Verlag, Berlin, 2007.

[4] P. Cannarsa, P. Martinez, and J. Vancostenoble. Persistent regional null controllability for a class of degenerate parabolic equations. *Commun. Pure Appl. Anal.*, 3(4):607–35, 2004.

[5] P. Cannarsa, P. Martinez, and J. Vancostenoble. Carleman estimates for a class of degenerate parabolic operators. *SIAM J. Control Optim.*, 47(1):1–19, 2008.

[6] P. Cannarsa, P. Martinez, and J. Vancostenoble. Carleman estimates and null controllability for boundary-degenerate parabolic operators. *C. R. Math. Acad. Sci. Paris*, 347(3–4):147–52, 2009.

[7] P. Cannarsa, P. Martinez, and J. Vancostenoble. The cost of controlling degenerate parabolic equations by boundary controls. *ArXiv e-prints*, arXiv:1511.06857, 2015.

[8] C. Cazacu. Controllability of the heat equation with an inverse-square potential localized on the boundary. *SIAM J. Control Optim.*, 52(4):2055–89, 2014.

[9] D. Colton. Integral operators and reflection principles for parabolic equations in one space variable. *J. Diff. Eq.*, 15:551–9, 1974.

[10] S. Ervedoza. Control and stabilization properties for a singular heat equation with an inverse-square potential. *Comm. Partial Diff. Eq.*, 33(10–12):1996–2019, 2008.

[11] H. Fattorini and D. Russell. Exact controllability theorems for linear parabolic equations in one space dimension. *Arch. Rational Mech. Anal.*, 43(4):272–92, 1971.

[12] C. Flores and L. de Teresa. Carleman estimates for degenerate parabolic equations with first order terms and applications. *C. R. Math. Acad. Sci. Paris*, 348(7–8):391–6, 2010.

[13] A. V. Fursikov and O. Y. Imanuvilov. *Controllability of Evolution Equations*, volume 34 of Lecture Notes Series. Seoul National University Research Institute of Mathematics Global Analysis Research Center, 1996.

[14] O. Y. Imanuvilov. Controllability of parabolic equations. *Mat. Sb.*, 186(6):109–32, 1995.

[15] B. Jones Jr. A fundamental solution for the heat equation which is supported in a strip. *J. Math. Anal. Appl.*, 60(2):314–24, 1977.

[16] B. Laroche. Extension de la notion de platitude à des systèmes décrits par des équations aux dérivées partielles linéaires. PhD thesis, Ecole des Mines de Paris, 2000.

[17] B. Laroche and P. Martin. Motion planning for a 1-D diffusion equation using a Brunovsky-like decomposition. In *Proceedings of the International Symposium on the Mathematical Theory of Networks and Systems (MTNS)*, Perpignan, France, 2000.

[18] B. Laroche, P. Martin, and P. Rouchon. Motion planning for the heat equation. *Int. J. Robust Nonlinear Control*, 10(8):629–43, 2000.

[19] G. Lebeau and L. Robbiano. Contrôle exact de l'équation de la chaleur. *Comm. Partial Diff. Eq.*, 20(1–2):335–56, 1995.

[20] W. A. J. Luxemburg and J. Korevaar. Entire functions and Müntz–Szász type approximation. *Trans. Amer. Math. Soc.*, 157:23–37, 1971.

178   *Philippe Martin, Lionel Rosier, and Pierre Rouchon*

[21] P. Martin, L. Rosier, and P. Rouchon. Null controllability of the 1D heat equation using flatness. In *1st IFAC workshop on Control of Systems Governed by Partial Differential Equations (CPDE2013)*, Institut Henri Poincaré, Paris, France, pp. 7–12, 2013.

[22] P. Martin, L. Rosier, and P. Rouchon. Null controllability of the 2D heat equation using flatness. In *52nd IEEE Conference on Decision and Control (CDC 2013)*, Florence, Italy, pp. 3738–43, 2013.

[23] P. Martin, L. Rosier, and P. Rouchon. Controllability of the 1D Schrödinger equation by the flatness approach. In *19th World Congress of the International Federation of Automatic Control (IFAC 2014)*, Cape Town, South Africa, pp. 646–51, 2014.

[24] P. Martin, L. Rosier, and P. Rouchon. Null controllability of the heat equation using flatness. *Autom. J. IFAC*, 50(12):3067–76, 2014.

[25] P. Martin, L. Rosier, and P. Rouchon. Null controllability using flatness: A case study of 1D heat equation with discontinuous coefficients. In *European Control Conference ECC15*, Linz, Austria, 2015.

[26] P. Martin, L. Rosier, and P. Rouchon. Flatness and null controllability of 1D parabolic equations. *PAMM*, 16(1):47–50, 2016.

[27] P. Martin, L. Rosier, and P. Rouchon. Null controllability of one-dimensional parabolic equations by the flatness approach. *SIAM J. Control Optim.*, 54(1):198–220, 2016.

[28] P. Martin, L. Rosier, and P. Rouchon. On the reachable states for the boundary control of the heat equation. *Appl. Math. Res. Express. AMRX*, 2016(2):181–216, 2016.

[29] T. Meurer. *Control of Higher-Dimensional PDEs: Flatness and Backstepping Designs*. Communications and Control Engineering. Springer, Berlin-Heidelberg, Germany, 2012.

CENTRE AUTOMATIQUE ET SYSTÈMES, MINES PARISTECH, PSL RESEARCH UNIVERSITY, 60 BOULEVARD SAINT-MICHEL, 75272 PARIS CEDEX 06, FRANCE
E-mail address: philippe.martin@mines-paristech.fr

CENTRE AUTOMATIQUE ET SYSTÈMES, MINES PARISTECH, PSL RESEARCH UNIVERSITY, 60 BOULEVARD SAINT-MICHEL, 75272 PARIS CEDEX 06, FRANCE
E-mail address: lionel.rosier@mines-paristech.fr

CENTRE AUTOMATIQUE ET SYSTÈMES, MINES PARISTECH, PSL RESEARCH UNIVERSITY, 60 BOULEVARD SAINT-MICHEL, 75272 PARIS CEDEX 06, FRANCE
E-mail address: pierre.rouchon@mines-paristech.fr

9

Mixing for the Burgers Equation Driven by a Localized Two-Dimensional Stochastic Forcing

ARMEN SHIRIKYAN

Abstract

We consider the one-dimensional Burgers equation perturbed by a stochastic forcing, which is assumed to be white in time and localized and low-dimensional in space. We establish a mixing property for the Markov process associated with the problem in question. The proof is based on a general criterion for mixing and a recent result on global approximate controllability to trajectories for damped conservation laws.

1991 Mathematics Subject Classification. 35K10, 35R60, 60H15

Key words and phrases. Stochastic Burgers equation, mixing, localized degenerate random perturbation

This research was supported by the RSF grant 14-49-00079

Contents

9.1 Introduction

The paper is devoted to studying the problem of uniqueness and mixing of a stationary measure for the following stochastic Burgers equation:

$$\partial_t u - \nu \partial_x^2 u + u \partial_x u = h(x) + \sum_{j=1}^{2} \dot{\beta}_j(t) e_j(x), \quad x \in I, \qquad (9.1)$$

$$u(t, 0) = u(t, \pi) = 0, \qquad (9.2)$$

where $I = (0, \pi)$, $\nu > 0$ is a parameter, $h \in L^\infty(I)$ is a fixed function, $\{\beta_j\}$ are independent Brownian motions, and $\{e_j\}$ are continuous functions supported by an interval $[a, b] \subset I$. The Cauchy problem for (9.1) and (9.2) is well posed (see chapter 14 in [5]): for any $u_0 \in L^2(I)$ there is a unique stochastic process $u(t, x)$ whose almost every trajectory coincides with u_0 for $t = 0$ and satisfies the inclusion

$$u \in \mathcal{X} := C(\mathbb{R}_+, L^2(I)) \cap L^2_{\text{loc}}(\mathbb{R}_+, H^1_0(I)) \qquad (9.3)$$

and (9.1) in the sense of distributions (see the notation section below for the definition of functional spaces). Moreover, the family of all solutions form a Markov process in $L^2(I)$, which possesses the Feller property. The latter holds also in the more regular space $H^1_0(I)$. The Bogolyubov–Krylov argument enables one to prove the existence of a stationary measure (see chapter 14 in [5]). Our aim is to study the uniqueness and mixing of a stationary measure.

The main result of this article proves that, for a particular choice of the functions e_1 and e_2, the Markov process associated with problem (9.1) and (9.2) has a unique stationary measure μ_ν for any $\nu > 0$, and the law of any solution converges weakly to μ_ν as $t \to +\infty$ in the Kantorovich–Wasserstein metric. An exact formulation of this result is given in Section 9.2.1, and the proof is presented in Section 9.3.

Let us mention that the problem of mixing of dissipative PDEs with a random external force was the focus of the attention of many researchers over the last 20 years, and it is rather well understood in the case when all deterministic modes are forced (see the literature review in chapter 3 of [13]). The situation in which the support of the random forcing does not cover all deterministic modes is far less studied, and only a few results are available. Namely, Hairer and Mattingly [9, 10] established the uniqueness of a stationary measure and its exponential stability for 2D Navier–Stokes equations with a random forcing, white in time and finite-dimensional in space, provided that the equation is considered on a torus or a sphere, and the unperturbed problem has a unique globally stable equilibrium. Bortichev [2] considered the viscous Burgers equation on a circle and proved that, if the unperturbed dynamics is globally asymptotically stable, then for any random perturbation, the stationary measure is unique, and convergence to it holds with a polynomial rate independent of the viscosity; see also [3, 6, 12] for the case of an inviscid equation. The paper [17] establishes the property of exponential mixing for the 2D Navier–Stokes system in a bounded domain in the case when the noise is smooth, bounded, and localized in the physical space and in time. Finally, Földes et al. [7] proved that the Hairer–Mattingly result is true for the Boussinesq system perturbed by a highly-degenerate stochastic forcing which acts only on the equation for temperature. To the best of our knowledge, the situation in which the unperturbed dynamics is nontrivial while the noise is low-dimensional and localized in space was not considered earlier, and this is the main novelty of this work. At the same time, it remains an open question to determine the rate of convergence to the stationary measure.

9.1.1 Notation

For a separable Banach space X, we denote by $\mathcal{B}_X$ the Borel σ-algebra on X, by $C_b(X)$ the space of bounded continuous functions $f: X \to \mathbb{R}$ endowed with the supremum norm $\| \cdot \|_\infty$, and by $L_b(X)$ the subspace of functions $f \in C_b(X)$ such that

$$\|f\|_L := \|f\|_\infty + \sup_{u \neq v} \frac{|f(u) - f(v)|}{\|u - v\|_X} < \infty.$$

The set of Borel probability measures on X is denoted by $\mathcal{P}(X)$ and endowed with the Kantorovich–Wasserstein distance (which metrizes the weak convergence of measures) defined as

$$\|\mu - \nu\|_L^* = \sup_{\|f\|_L \leq 1} |\langle f, \mu \rangle - \langle f, \nu \rangle|,$$

where $\langle f, \mu \rangle$ stands for the integral of f over X with respect to μ. Given $R > 0$, we write $B_X(R)$ for the closed ball in X centered at zero of radius R.

Let $I = (0, \pi)$ and let $J \subset \mathbb{R}$ be a closed interval. We use the following functional spaces.

$C(J, X)$ is the space of continuous functions $f: J \to X$.

$L^2(J, X)$ is the space of Borel-measurable functions $f: J \to X$ such that $\|f(t)\|_X$ is square-integrable on J. If J is unbounded, then $L^2_{\text{loc}}(J, X)$ stands for the space of X-valued functions on J whose restriction to any finite interval $J' \subset J$ belongs to $L^2(J', X)$.

$H^s(I)$ is the Sobolev space of order $s \geq 0$ with a standard norm $\| \cdot \|_s$ and $H_0^s(I)$ is the closure in $H^s(I)$ of the space of infinitely smooth functions with compact support. We write $H = L^2(I)$, $V = H_0^1(I)$, $U = H^2(I) \cap V$, and denote by $\|u\|$ and $\|u\|_V = \|\partial_x u\|$ the norms in H and V, respectively.

$\mathcal{L}(X, Y)$ stands for the space of continuous linear operators from X to Y. We denote by C_i, $i = 1, 2, \ldots$, unessential positive numbers.

9.2 Main Result

9.2.1 The Result

We consider problem (9.1) and (9.2) on the interval $I = (0, \pi)$. Under the regularity hypotheses on e_1, e_2, and h mentioned in the Introduction, this problem is globally well posed and generates a Feller family of Markov processes, which will be denoted by $(u_t, \mathbb{P}_u)$. Let $P_t(u, \Gamma) = \mathbb{P}_u\{u_t \in \Gamma\}$ be its transition function and let

$$\mathfrak{P}_t : C_b(H) \to C_b(H), \quad \mathfrak{P}_t^* : \mathcal{P}(H) \to \mathcal{P}(H)$$

be the corresponding Markov semigroups. Recall that a measure $\mu \in \mathcal{P}(H)$ is said to be *stationary* for $(u_t, \mathbb{P}_u)$ if $\mathfrak{P}_t^* \mu = \mu$ for $t \geq 0$.

Let us fix an interval $[a, b] \subset I$ and assume that the functions e_j entering (9.1) have form

$$e_1(x) = c_1 \sin\left(\pi \frac{x - a}{b - a}\right), \quad e_2(x) = c_2 \sin\left(2\pi \frac{x - a}{b - a}\right) \quad \text{for } x \in [a, b],$$

where $c_1, c_2 \in \mathbb{R}$ are nonzero numbers, and $e_1(x) = e_2(x) = 0$ for $x \notin [a, b]$. The following theorem is the main result of this chapter.

Theorem 9.1 *For any $\nu > 0$, the Markov family $(u_t, \mathbb{P}_u)$ associated with (9.1) and (9.2) has a unique stationary measure $\mu_\nu \in \mathcal{P}(H)$. Moreover, the measure μ_ν is concentrated on V, and there is a function $\alpha(t)$, defined on the positive half-line and going to zero as $t \to +\infty$, such that*

$$\|\mathfrak{P}_t^* \lambda - \mu_\nu\|_L^* \leq \alpha(t) \quad \text{for } t \geq 1, \tag{9.4}$$

where $\lambda \in \mathcal{P}(H)$ is an arbitrary initial measure, and $\| \cdot \|_L^*$ is the Kantorovich–Wasserstein distance considered over the space V.

Let us emphasize that the convergence in (9.4) (called *mixing property* for μ_ν) holds uniformly with respect to the initial measures λ. This is due to the strong dissipative character of the Burgers equation (cf. [2]).

A proof of Theorem 9.1 is given in the next section. Here we present the general idea of the proof and outline the main steps.

9.2.2 Scheme of the Proof of Theorem 9.1

As was mentioned in the Introduction, the existence of a stationary measure μ_ν follows from the Bogolyubov–Krylov argument and standard a priori estimates for solutions of the Burgers equation (e.g., see section 2.5 in [13] for the more complicated case of the Navier–Stokes system). Thus, we shall concentrate on the proof of uniqueness of μ_ν and mixing property.

Step 1. Reduction to regular solutions. The Burgers equation possesses the following regularizing property:

Regularization For any $\lambda \in \mathcal{P}(H)$ and $t > 0$, we have $(\mathfrak{P}_t^*\lambda)(V) = 1$.

It follows that any stationary measure is concentrated on V, and it suffices to prove uniqueness of stationary distribution in the class of probability measures on V. Furthermore, when proving convergence (9.4), one can assume that $\lambda(V) = 1$. Finally, recalling the relation

$$(\mathfrak{P}_t^*\lambda)(\Gamma) = \int_V P_t(v,\Gamma)\lambda(dv), \quad \Gamma \in \mathcal{B}_H,$$

we see that it suffices to prove (9.4) with $\lambda = \delta_v$ with an arbitrary $v \in V$:

$$\|P_t(v,\cdot) - \mu_\nu\|_L^* \le \alpha(t) \quad \text{for } t \ge 1, v \in V. \tag{9.5}$$

Step 2. A sufficient condition for uniform mixing. We shall consider the restriction of the Markov process $(u_t, \mathbb{P}_u)$ to the space V. Let us define the Markov process $(\boldsymbol{u}_t, \mathbb{P}_{\boldsymbol{u}})$ in $\boldsymbol{V} = V \times V$ as a pair of independent copies of $(u_t, \mathbb{P}_u)$. In other words, the probability space is defined as the direct product $\boldsymbol{\Omega} = \Omega \times \Omega$, with the natural product σ-algebra on it, the process $\boldsymbol{u}_t$ is given by

$$\boldsymbol{u}_t(\boldsymbol{\omega}) = \big(u_t(\omega), u_t(\omega')\big), \quad \boldsymbol{\omega} = (\omega,\omega') \in \boldsymbol{\Omega},$$

and $\mathbb{P}_{\boldsymbol{u}} = \mathbb{P}_u \otimes \mathbb{P}_{u'}$ for $\boldsymbol{u} = (u, u')$. In view of theorem 3.1.3 in [13] and the remark following it, the uniqueness of a stationary measure and convergence (9.5) will be established if we prove the following two properties:

Stability There is a sequence $\{B_m\}$ of closed subsets of V such that

$$\sup_{t \geq 1} \|P_t(v, \cdot) - P_t(v', \cdot)\|_L^* \leq \delta_m \quad \text{for } v, v' \in B_m, \qquad (9.6)$$

where $\{\delta_m\}$ is a sequence of positive numbers going to zero as $m \to \infty$, and $\|\cdot\|_L^*$ stands for the Kantorovich–Wasserstein distance over V.

Recurrence Let τ_m be the first hitting time of the set $\boldsymbol{B}_m = B_m \times B_m$ for the process $(\boldsymbol{u}_t, \mathbb{P}_{\boldsymbol{u}})$. Then

$$\sup_{\boldsymbol{u} \in \boldsymbol{V}} \mathbb{P}_{\boldsymbol{u}}\{\tau_m > t\} \leq p(m, t) \quad \text{for any } m \geq 1, t > 0, \qquad (9.7)$$

where $p(m, t) \to 0$ as $t \to +\infty$.

The proofs of these assertions (outlined below) are based on the contraction of the L^1-norm of the difference between two solutions of the Burgers equation, a strong dissipation property and parabolic regularization, and a recent controllability result for damped-driven conservation laws.

Step 3. Stability. It is a well-known fact that, for any initial functions $v, v' \in V$, the corresponding solutions of (9.1) and (9.2) satisfy the inequality

$$\|u(t) - u'(t)\|_{L^1(I)} \leq \|v - v'\|_{L^1(I)} \quad \text{for } t \geq 0; \qquad (9.8)$$

(see section 3.3 in [11]). On the other hand, the a priori estimates for solutions of (9.1) and (9.2) imply that, if an initial function belongs to a bounded set $B \subset V$, then

$$\sup_{t \geq 1} \mathbb{E} \|u(t)\|_{H^s} \leq C_s(B), \qquad (9.9)$$

where $s \in (1, 2)$ is arbitrary. Combining inequalities (9.8) and (9.9) with an interpolation inequality, we can easily prove (9.6) with any sequence of bounded sets $B_m \subset V$ whose diameters in L^1 go to zero.

Step 4. Recurrence. A comparison theorem (see section 2.2 in [1]) and regularization for parabolic equations imply that the solutions of (9.1) and (9.2) satisfy the inequality

$$\mathbb{P}_v\{\|u_2\|_s \leq R\} \geq p \quad \text{for } v \in V, \qquad (9.10)$$

where $s \in (1, 2)$ is fixed, $p > 0$ is a number not depending on ν, and $R > 0$ is sufficiently large. On the other hand, using global controllability to trajectories of the Burgers equation (see theorem 1.2 in [16]) and the continuous dependence of trajectories on the noise, one can prove that

$$\mathbb{P}_\nu \{ u_{T_m} \in B_m \} \geq \varepsilon_m > 0 \quad \text{for } \nu \in B_{V \cap H^s}(R), \tag{9.11}$$

where $B_m \subset V$ are bounded closed sets whose L^1-diameters go to zero as $m \to \infty$, and $T_m > 0$ are some numbers. Combining (9.10) and (9.11) with the Markov property, we derive the required result.

9.3 Proof of the Main Result

As was explained in Section 9.2.2, Theorem 9.1 will be established if we prove the regularization, stability, and recurrence properties stated in Steps 1 and 2. This is done in the next three subsections.

9.3.1 Regularization

We wish to prove that, if $u(t, x)$ is the solution of (9.1) and (9.2) issued from an initial point $u_0(x)$, which is an H-valued random variables independent of (β_1, β_2), then

$$\mathbb{P}\{u(t) \in V\} = 1 \quad \text{for any } t > 0. \tag{9.12}$$

We begin with a number of simple remarks. In view of the relation

$$\mathbb{P}\{u(t) \in V\} = \int_H P_t(z, V) \lambda(dz),$$

where λ is the law of the initial state u_0, it suffices to consider the case in which u_0 is a deterministic function. Applying the Ito formula to the L^2 norm of a solution, using the L^2-orthogonality of the nonlinear term and solution, and repeating standard arguments (e.g., see the proof of proposition 2.4.8 in [13]), we derive

$$\mathbb{E}\|u(t)\|^2 + \mathbb{E}\int_0^t \|u(r)\|_V^2 \, dr \leq C_1(1 + t) \quad \text{for all } t \geq 0. \tag{9.13}$$

It follows that $\mathbb{P}\{u(r) \in V\} = 1$ for almost every $r \in (0, t)$. Thus, when proving (9.12), we can assume from the very beginning that u_0 is a deterministic function belonging to V.

Let us denote by $z(t,x)$ the solution of the problem

$$\partial_t z - \nu \partial_x^2 z = \eta(t,x), \quad z(t,0) = z(t,\pi) = 0, \quad z(0,x) = 0, \qquad (9.14)$$

where η stands for the stochastic term in (9.1), and represent a solution of (9.1)–(9.3) in the form $u = z + v$. In this case, $v(t,x)$ must be a solution of the problem

$$\partial_t v - \nu \partial_x^2 v + (v+z)\partial_x(v+z) = h(x), \qquad (9.15)$$
$$v(t,0) = v(t,\pi) = 0, \qquad (9.16)$$
$$v(0,x) = u_0(x). \qquad (9.17)$$

The following lemma can be established by well-known arguments, and its proof is outlined at the end of this section.

Lemma 9.2 Almost every trajectory of z belongs to the space

$$\mathcal{Y}_s := C(\mathbb{R}_+, V \cap H^s) \cap L^2_{\text{loc}}(\mathbb{R}_+, U), \quad s < 2,$$

and there are positive numbers $C_k = C_k(\nu)$ such that

$$\sup_{t \geq 0} \mathbb{E}\left(\|z(t)\|_1^2 + \int_t^{t+1} \|z(r)\|_2^2 \, dr \right)^k \leq C_k. \qquad (9.18)$$

Moreover, for any $\nu > 0$, $k \geq 1$, $T > 0$, and $s \in (1,2)$, there is $C = C(k,T,s,\nu) > 0$ such that

$$\mathbb{E} \sup_{t \leq r \leq t+T} \|z(r)\|_s^k \leq C \quad \text{for any } t \geq 0. \qquad (9.19)$$

In view of the continuous embedding $H^s \subset C(\bar{I})$ valid for $s > 1/2$ and inequality (9.19), for any $T > 0$, the quantity

$$R_T := \sup_{0 \leq t \leq T} \left(\|z(t,\cdot)\|_{L^\infty} + \|\partial_x z(t,\cdot)\|_{L^\infty} \right) \qquad (9.20)$$

is almost surely finite. Applying the maximum principle to (9.15)–(9.17), (e.g., see section 2 in [14, chapter 3]), we derive

$$\|v\|_{L^\infty(D_T)} \leq C_2(R_T + 1) + \|u_0\|_{L^\infty}, \qquad (9.21)$$

where we set $D_T = (0,T) \times I$. It follows that

$$\|u\|_{L^\infty(D_T)} = \|v+z\|_{L^\infty(D_T)} \leq (C_2 + 1)(R_T + 1) + \|u_0\|_{L^\infty}. \qquad (9.22)$$

Combining this with (9.13), we see that

$$g := u\partial_x u - h = (v+z)\partial_x(v+z) - h \in L^2(D_T) \quad \text{almost surely.} \qquad (9.23)$$

We now use the Duhamel formula to write a solution of (9.15)–(9.17) in the form

$$v(t, x) = (e^{tL}u_0)(x) - \int_0^t e^{(t-r)L}g(r, \cdot)\, dr, \tag{9.24}$$

where $L = \nu \partial_x^2$. Since the operator e^{tL} is continuous from H to V for $t > 0$, and its norm satisfies the inequality $\|e^{tL}\|_{\mathcal{L}(H,V)} \leq C_3 t^{-1/2}$, we see that, with probability 1, $v(t)$ is a continuous function of $t > 0$ with range in V. Recalling that, by Lemma 9.2, the same is true for $z(t)$, we arrive at (9.12).

9.3.2 Stability

We first show that inequalities (9.8) and (9.9) imply the stability property with any sequence of closed sets $B_m \subset V$ such that

$$\sup_{v \in B_m} \|v\|_V \leq C_4 \quad \text{for any } m \geq 1, \qquad \sup_{v, v' \in B_m} \|v - v'\|_{L^1} \to 0 \quad \text{as } m \to \infty, \tag{9.25}$$

where $C_4 > 0$ does not depend on m. Indeed, given $v, v' \in B_m$, we denote by $u(t)$ and $u'(t)$ the solutions of (9.1)–(9.3) with $u_0 = v$ and $u_0 = v'$, respectively. If $f \in L_b(V)$ is a function with $\|f\|_L \leq 1$, then

$$\left| \langle f, P_t(v, \cdot) \rangle - \langle f, P_t(v', \cdot) \rangle \right| = \left| \mathbb{E}\big(f(u(t)) - f(u'(t))\big) \right| \leq \mathbb{E}\|u(t) - u'(t)\|_V. \tag{9.26}$$

Recall now that, for any $s \in [1, 2]$, we have the interpolation inequality (e.g., see section 1.6 in [8])

$$\|u\|_V \leq C_5 \|u\|_{L^1}^{1-\theta_s} \|u\|_s^{\theta_s}, \quad u \in H^s,$$

where we set $\theta_s = \frac{3}{2s+1}$. Combining this with (9.26) and taking the supremum over all $f \in L_b(V)$ with $\|f\|_L \leq 1$, we derive

$$\|P_t(v, \cdot) - P_t(v', \cdot)\|_L^* \leq \mathbb{E}\left\{ \|u(t) - u'(t)\|_{L^1}^{1-\theta_s} \big(\|u(t)\|_s + \|u'(t)\|_s\big)^{\theta_s} \right\}.$$

It follows from (9.8) and (9.9) that, for $t \geq 1$, the right-hand side of this inequality does not exceed $C_6 \|v - v'\|_{L^1}^{1-\theta_s}$, where $C_6 > 0$ depends only on the H^1 norm of the initial functions. It is now straightforward to see that, if (9.25) is satisfied, then (9.6) holds with a sequence δ_m going to zero as $m \to \infty$.

Thus, to complete the proof of stability, it remains to establish (9.9). In view of Lemma 9.2, it suffices to prove that inequality (9.9) is true for the solution $v(t, x)$ of problem (9.15)–(9.17). To this end, we repeat essentially the argument of Section 9.3.1, using the relation (cf. (9.24))

$$v(t, x) = e^L v(t-1) - \int_{t-1}^t e^{(t-r)L} g(r, \cdot)\, dr, \qquad (9.27)$$

where g is defined by (9.23), as well as the inequality

$$\|e^{tL}\|_{\mathcal{L}(H, H^s)} \leq C_7 t^{-s/2} \quad \text{for } t \geq 0, \qquad (9.28)$$

where $s \in [1, 2]$. Namely, suppose we have proved that

$$\mathbb{E} \sup_{t \leq r \leq t+1} \|v(r)\|_1^2 \leq C_8 \quad \text{for } t \geq 0. \qquad (9.29)$$

Combining this with (9.28), we obtain the following estimate for the first term in (9.27):

$$\mathbb{E} \|e^L v(t-1)\|_2 \leq C_9 \quad \text{for all } t \geq 1. \qquad (9.30)$$

Denoting by $G(t)$ the second term in (9.27) and using again (9.28), we write

$$\|G(t)\|_s \leq C_7 \int_{t-1}^t (t-r)^{-s/2} \|g(r)\|\, dr \leq C_{10} \sup_{t-1 \leq r \leq t} \|g(r)\|. \qquad (9.31)$$

It follows from the definition of g that (see (9.23))

$$\|g(r)\| \leq C_{11}\left(\|z(r)\|_1^2 + \|v(r)\|_1^2\right) + \|h\|.$$

Combining this with (9.19), (9.29), and (9.31), we obtain $\mathbb{E}\|G(t)\|_s \leq C_{12}$ for $t \geq 1$. Recalling (9.27) and (9.30), we arrive at inequality (9.9) with u replaced by v.

Thus, it remains to establish (9.29). Without loss of generality, we shall assume that $t \geq 1$. In view of inequalities (9.13) and (9.18), we can find $t_0 \in [t-1, t]$ such that

$$\mathbb{E} \|v(t_0)\|_1^2 \leq C_{13}. \qquad (9.32)$$

Furthermore, standard argument (see proposition 2.4.9 in [13]) combined with inequality (9.18) shows that

$$\mathbb{E} \|v(t)\|^k \leq C_{14}(k) \quad \text{for } t \geq 0, \ k \geq 1. \qquad (9.33)$$

We now take the scalar product in L^2 of (9.15) with $-2\partial_x^2 v$. Integrating by parts and using the Hölder inequality, we get

$$\partial_t \|\partial_x v\|^2 + 2\nu \|\partial_x^2 v\|^2 \leq 2\left((z+v)\partial_x(z+v) - h, \partial_x^2 v\right)$$
$$\leq \nu \|\partial_x^2 v\|^2 + 2\|\partial_x v\|_{L^3}^3 + C_{15} R(t), \qquad (9.34)$$

where $(\cdot, \cdot)$ denotes the scalar product in L^2, $R(t) = 1 + \|z(t)\|_\sigma^4$, and $\sigma \in (\frac{3}{2}, 2)$ is a fixed number. Using the continuous embedding $L^3 \subset H^{1/6}$ and an interpolation inequality for Sobolev spaces, we derive

$$2\|\partial_x v\|_{L^3}^3 \le C_{16}\|v\|_{7/6}^3 \le C_{17}\|v\|^{5/4}\|\partial_x^2 v\|^{7/4} \le \nu\|\partial_x^2 v\|^2 + C_{18}\|v\|^{10}.$$

Substituting this into (9.34) and integrating in time, we get

$$\sup_{t \le r \le t+1} \|\partial_x v(r)\|^2 \le \|\partial_x v(t_0)\|^2 + C_{18} \int_{t_0}^{t+1} \left(\|v(r)\|^{10} + R(r) \right) dr.$$

Taking the mean value and using (9.19), (9.32), and (9.33), we arrive at (9.29). This completes the proof of the stability property.

9.3.3 Recurrence

Inequalities (9.10) and (9.11) combined with the independence of the components of the extended process $(\boldsymbol{u}_t, \mathbb{P}_{\boldsymbol{u}})$ imply that

$$\mathbb{P}_{\boldsymbol{v}}\{\|u_2\|_s \le R, \|u_2'\|_s \le R\} \ge p^2 \quad \text{for } v, v' \in V,$$
$$\mathbb{P}_{\boldsymbol{v}}\{(u_{T_m}, u_{T_m}') \in \boldsymbol{B}_m\} \ge \varepsilon_m^2 \quad \text{for } v, v' \in B_{V \cap H^s}(R),$$

where $\boldsymbol{v} = (v, v')$. It is a standard fact in the theory Markov processes that (9.7) follows from these two inequalities (e.g., see the proof of proposition 3.3.6 in [13]). We thus confine ourselves to the proof of (9.10) and (9.11).

Proof of (9.10) The Kolmogorov–Chapman relation implies that it suffices to establish the following two inequalities:

$$P_1(v, B_{L^\infty}(K)) \ge p_1 \quad \text{for any } v \in V, \tag{9.35}$$
$$P_1(v, B_{V \cap H^s}(R)) \ge \tfrac{1}{2} \quad \text{for any } v \in B_{L^\infty}(K), \tag{9.36}$$

where $s \in (1, 2)$ is any fixed number, $p_1 > 0$ does not depend on v, and K and R are sufficiently large positive numbers. The proofs of these inequalities are based on a comparison principle and a regularization property for parabolic equations.

To prove (9.35), we write $u = z + v$ and note that, in view of (9.18),

$$\mathbb{P}\{\|z(1)\|_{L^\infty} \le R\} \to 1 \quad \text{as } R \to \infty.$$

Therefore it suffices to find $R_1 > 0$ such that

$$\mathbb{P}\{\|v(1)\|_{L^\infty} \le R_1\} \ge p_2 \quad \text{for any } u_0 \in V, \tag{9.37}$$

where $p_2 > 0$ does not depend on u_0, and $v(t, x)$ stands for the solution of (9.15)–(9.17). Let us fix any $\sigma \in (\frac{3}{2}, 2)$ and, given $\rho > 0$, define the event

$$\Gamma_\rho = \left\{ \sup_{0 \leq t \leq 1} \|z(t)\|_\sigma \leq \rho \right\}.$$

Since z is a zero-mean Gaussian process in V with Hölder continuous trajectories, there is a positive function $p(\rho)$ of the variable $\rho > 0$ such that $\mathbb{P}(\Gamma_\rho) \geq p(\rho)$. We claim that, if $\rho > 0$ is sufficiently small and $R_1 > 0$ is sufficiently large, then

$$\|v(1)\|_{L^\infty} \leq R_1 \quad \text{on the event } \Gamma_\rho. \tag{9.38}$$

Once this property is established, the required inequality (9.37) will follow from the lower bound for $\mathbb{P}(\Gamma_\rho)$.

The proof of (9.38) is based on a comparison principle for parabolic equations. We first show that

$$v(1, x) \leq R_1 \quad \text{for all } x \in I. \tag{9.39}$$

To this end, we shall construct a supersolution $v_+(t, x)$ for problem (9.15)–(9.17) in the domain $D_1 = (0, 1) \times I$, that is, a smooth function which is positive on the lateral boundary of D_1 and greater than u_0 for $t = 0$, and satisfies the inequality

$$\partial_t v_+ - \nu \partial_x^2 v_+ + (v_+ + z)\partial_x(v_+ + z) - h(x) \geq 0 \quad \text{for } (t, x) \in D_1. \tag{9.40}$$

Let us set

$$v_+(t, x) = \frac{\delta(x + M) + C\varepsilon}{t + \varepsilon},$$

where $C > \|u_0\|_{L^\infty}$ is fixed, ε and δ are small positive parameters, and $M > 0$ is a large parameter. It is straightforward to check that $v_+(t, 0) > 0$ and $v_+(t, \pi) > 0$ for $t \in [0, 1]$ and $v_+(0, x) > \|u_0\|_{L^\infty}$ for $x \in \bar{I}$. Furthermore, it is a matter of a simple computation to show that, if $\varepsilon \in (0, 1)$ and $\rho > 0$ is sufficiently small, then (9.40) holds, provided that $\delta > 0$ is small (for instance, one can take $\delta = \frac{1}{2}$) and $M > 0$ sufficiently large. Therefore, by theorem 2.2 in [1], we have

$$v(t, x) \leq v_+(t, x) = \frac{\delta(x + M) + C\varepsilon}{t + \varepsilon} \quad \text{for } (t, x) \in D_1.$$

Taking $t = 1$ and letting $\varepsilon \to 0^+$, we see that (9.39) holds with $R_1 = \delta(\pi + M)$. A similar argument shows that the function $v_- = -v_+$ is a subsolution for (9.15)–(9.17) in D_1, and therefore $v(1, x) \geq -R_1$ for $x \in I$.

It remains to prove (9.36). If the set of initial functions u_0 was bounded in B, inequality (9.9) would imply the required result. However, in our case, it is bounded in L^∞. To overcome this difficulty, let us remark that, in view of (9.13), we can find $t_0 \in (0, \frac{1}{2})$ such that $\mathbb{E}\|u(t_0)\|_V^2 \le 3C_1$. It follows that, if B is a ball in V centered at zero of sufficiently large radius, then

$$\mathbb{P}_{u_0}\{u(t_0) \in B\} \ge 2^{-1/2} \quad \text{for any } u_0 \in B_{L^\infty}(K).$$

The Kolmogorov–Chapman relation and an analogue of inequality (9.9) on the half-line $t \ge \frac{1}{2}$ now implies the required result.

Proof of (9.11) Along with (9.1), let us consider the controlled equation

$$\partial_t u - \nu \partial_x^2 u + u \partial_x u = h(x) + \zeta(t, x), \quad x \in (0, \pi), \qquad (9.41)$$

where $\zeta(t)$ is a control function with range in $E := \text{span}\{e_1, e_2\}$. Let $\hat{u}(x)$ be a time-independent solution of problem (9.41), (9.2) with $\zeta \equiv 0$. Such a solution exists and belong to $H^2 \cap V$ (e.g., see section 7 in [15, chapter I]). The following proposition is a consequence of the main result of [16].

Proposition 9.3 There is $M > 0$ such that, given $\varepsilon > 0$, one can find $T_\varepsilon > 0$ satisfying the following property: for any $u_0 \in V$ there exists a continuous function $\zeta : [0, T_\varepsilon] \to E$ such that

$$\|u(T_\varepsilon) - \hat{u}\|_{L^1} < \varepsilon, \quad \|u(T_\varepsilon)\|_V < M, \qquad (9.42)$$

where $u(t, x)$ stands for the solution of (9.2), (9.3), and (9.41).

A well-known argument based on the above result on approximate controllability (see the proof of theorem 7.4.1 in [5]) enables one to prove that

$$p_{\varepsilon, M}(v) := P_{T_\varepsilon}(v, B(\varepsilon, M)) > 0 \quad \text{for any } v \in V, \qquad (9.43)$$

where we set

$$B(\varepsilon, M) = \{u \in V : \|u - \hat{u}\|_{L^1} < 2\varepsilon, \|u\|_V < 2M\}.$$

Since $B(\varepsilon, M)$ is an open subset in V, the Feller property of the Markov process $(u_t, \mathbb{P}_u)$ in V implies that the function $p_{\varepsilon, M}(v)$ is lower semicontinuous and, hence, separated from zero on compact subsets. In particular, denoting by B_m the closure of $B(\frac{1}{m}, M)$ in V, we can find positive numbers $\varepsilon_m > 0$ such that

$$P_{T_m}(v, B_m) \ge p_{\frac{1}{m}, M}(v) \ge \varepsilon_m \quad \text{for } v \in B_{V \cap H^s}(R),$$

where $T_m = T_{\frac{1}{m}}$. It remains to note that the sets B_m satisfy (9.25) with $C_4 = 2M$. This completes the proof of the recurrence property.

9.3.4 Proof of Lemma 9.2

The fact that almost every trajectory belongs to $\mathcal{Y}_1$ and inequality (9.18) are well known (see proposition 2.4.10 and corollary 2.4.11 in [13] for the case of the 2D Navier–Stokes system). Thus, we shall prove that $z \in C(\mathbb{R}_+, H^s)$ with probability 1 and that (9.19) holds for any $t \geq 0$, $s < 2$, and $k \geq 1$.

To establish the required properties, we first assume that $t = 0$. Using the factorization method described in section 5.3 of [4], we write z in the form

$$z(t) = \int_0^t e^{(t-r)L} y_\alpha(r)(t-r)^{\alpha-1}\, dr, \tag{9.44}$$

where $\alpha \in (0, \frac{1}{2})$ is an arbitrary number, $L = \nu \partial_x^2$, and

$$y_\alpha(r) = \sum_{j=1}^{2} \int_0^r e^{(r-\theta)L} e_j(r-\theta)^{-\alpha} d\beta_j(\theta).$$

As is shown in the proof of theorem 5.9 in [4], $y_\alpha(r)$ is a Gaussian process satisfying the inequality

$$\mathbb{E} \int_0^T \|y_\alpha(r)\|_1^n dr \leq C_n(\alpha) T \quad \text{for any } T > 0 \text{ and } n \geq 1. \tag{9.45}$$

Now note that

$$\|e^{tL}\|_{\mathcal{L}(V, H^s)} \leq C t^{(1-s)/2} \quad \text{for } s \in [1, 2] \text{ and } t \geq 0. \tag{9.46}$$

Taking the H^s norm in (9.44), using (9.46), and applying the Young inequality, we obtain

$$\|z(t)\|_s \leq \int_0^t (t-r)^{\alpha-\frac{1+s}{2}} \|y_\alpha(r)\|_V\, dr \leq \int_0^t \left((t-r)^{p_m} + \|y_\alpha(r)\|_V^m\right) dr, \tag{9.47}$$

where $p_m = \frac{m}{m-1}(\alpha - \frac{1+s}{2})$, and the number $m \geq 1$ is chosen below. For a given $s \in (1, 2)$, we can find α satisfying the inequality $\frac{1}{2} > \alpha > \frac{s-1}{2}$. In this case, we have $\alpha - \frac{1+s}{2} > -1$, and one can choose m so large that $p_m > -1$. Combining this with (9.45) and (9.47), we arrive at (9.19). The continuity of z as a function of time with values in H^s can now be proven by a simple approximation argument.

To prove (9.19) in the general case, we can assume that $t = t_0 \geq 1$. Let us write

$$z(t) = e^{(t-t_0-1)L} z(t_0 - 1) + \sum_{j=1}^{2} \int_{t_0-1}^{t} e^{(t-r)L} e_j\, d\beta_j(r). \tag{9.48}$$

The sum on the right-hand side of this relation satisfies the same estimate on the interval $[t_0, t_0 + T]$ as the function defined by (9.44) on $[0, T + 1]$. Furthermore, in view of (9.46), we have the following estimate for the first term on the right-hand side of (9.48):

$$\sup_{t_0 \le t \le t_0 + T} \| e^{(t - t_0 - 1)L} z(t_0 - 1) \|_2 \le C \| z(t_0 - 1) \|_V. \qquad (9.49)$$

Recalling (9.18), we see that the mean value of any degree of the left-hand side of (9.49) is bounded by a constant not depending on $t_0 \ge 1$. This completes the proof of the lemma.

References

[1] H. W. Alt and S. Luckhaus, Quasilinear elliptic–parabolic differential equations, *Math. Z.* **183** (1983), no. 3, 311–41.

[2] A. Boritchev, Sharp estimates for turbulence in white-forced generalised Burgers equation, *Geom. Funct. Anal.* **23** (2013), no. 6, 1730–71.

[3] A. Debussche and J. Vovelle, Invariant measure of scalar first-order conservation laws with stochastic forcing, *Probab. Theory Related Fields* **163** (2015), no. 3–4, 575–611.

[4] G. Da Prato and J. Zabczyk, *Stochastic Equations in Infinite Dimensions*, Cambridge University Press, Cambridge, 1992.

[5] G. Da Prato and J. Zabczyk, *Ergodicity for Infinite Dimensional Systems*, Cambridge University Press, Cambridge, 1996.

[6] W. E, K. Khanin, A. Mazel, and Ya. Sinai, Invariant measures for Burgers equation with stochastic forcing, *Ann. Math.* (2) **151** (2000), no. 3, 877–960.

[7] J. Földes, N. Glatt-Holtz, G. Richards, and E. Thomann, Ergodic and mixing properties of the Boussinesq equations with a degenerate random forcing, *J. Funct. Anal.* **269** (2015), no. 8, 2427–504.

[8] D. Henry, *Geometric Theory of Semilinear Parabolic Equations*, Springer-Verlag, Berlin, 1981.

[9] M. Hairer and J. C. Mattingly, Ergodicity of the 2D Navier–Stokes equations with degenerate stochastic forcing, *Ann. Math.* (2) **164** (2006), no. 3, 993–1032.

[10] M. Hairer and J. C. Mattingly, A theory of hypoellipticity and unique ergodicity for semilinear stochastic PDEs, *Electron. J. Probab.* **16** (2011), no. 23, 658–738.

[11] L. Hörmander, *Lectures on Nonlinear Hyperbolic Differential Equations*, Springer-Verlag, Berlin, 1997.

[12] R. Iturriaga and K. Khanin, Burgers turbulence and random Lagrangian systems, *Comm. Math. Phys.* **232** (2003), no. 3, 377–428.

[13] S. Kuksin and A. Shirikyan, *Mathematics of Two-Dimensional Turbulence*, Cambridge University Press, Cambridge, 2012.

[14] E. M. Landis, *Second Order Equations of Elliptic and Parabolic Type*, American Mathematical Society, Providence, RI, 1998.

[15] J.-L. Lions, *Quelques Méthodes de Résolution des Problèmes aux Limites Non Linéaires*, Dunod, Paris, 1969.

[16] S. Rodrigues and A. Shirikyan, Global exponential stabilisation for a damped-driven conservation law with localised control and applications, in preparation.

[17] A. Shirikyan, Control and mixing for 2D Navier–Stokes equations with space-time localised noise, *Ann. Sci. Éc. Norm. Supér.* (4) **48** (2015), no. 2, 253–80.

UNIVERSITY OF CERGY-PONTOISE (FRANCE) AND NATIONAL RESEARCH UNIVERSITY MPEI, MOSCOW (RUSSIA)

E-mail address: Armen.Shirikyan@u-cergy.fr

Printed in the United States
by Baker & Taylor Publisher Services